AF601916

# Analysis of Agricultural Subsidies in India

## The Author

**Dr. Rajwinder Kaur** is born in Patiala district of Punjab (India). She has done B.Com, M.A. (Economics), M.Phil. (Economics) and Ph.D. (Economics) from Punjabi University Patiala. She is a serious research worker and her findings have been published in different journals. She has also presented research papers in national as well as international conferences. She has taught economics to MBA, M.Com., BBA, B.Com. and B.A. classes in various educational institutes and also in Lovely Professional University and at presently working as Assistant Professor and Head, Department of Economics in Guru Hargobind Sahib Khalsa Girls College, Karhali Sahib, Distt. Patiala (Punjab).

# Analysis of Agricultural Subsidies in India

**Dr. Rajwinder Kaur**

*Ph.D, M.Phil, B.Ed., M.A., B.Com*

*Head and Assistant Professor*
*Department of Economics*
*Guru Hargobind Sahib Khalsa Girls College*
*Karhali Sahib, Distt. Patiala, Punjab*

2016

Scholars World

*A Division of*

Astral International Pvt. Ltd.

New Delhi - 110 002

© 2016 AUTHOR

*Publisher's note:*

*Every possible effort has been made to ensure that the information contained in this book is accurate at the time of going to press, and the publisher and author cannot accept responsibility for any errors or omissions, however caused. No responsibility for loss or damage occasioned to any person acting, or refraining from action, as a result of the material in this publication can be accepted by the editor, the publisher or the author. The Publisher is not associated with any product or vendor mentioned in the book. The contents of this work are intended to further general scientific research, understanding and discussion only. Readers should consult with a specialist where appropriate.*

*Every effort has been made to trace the owners of copyright material used in this book, if any. The author and the publisher will be grateful for any omission brought to their notice for acknowledgement in the future editions of the book.*

*All Rights reserved under International Copyright Conventions. No part of this publication may be reproduced, stored in a retrieval system, or transmitted in any form or by any means, electronic, mechanical, photocopying, recording or otherwise without the prior written consent of the publisher and the copyright owner.*

**Cataloging in Publication Data--DK**
Courtesy: D.K. Agencies (P) Ltd. <docinfo@dkagencies.com>

**Rajwinder Kaur, author.**
Analysis of agricultural subsidies in India / author, Dr. Rajwinder Kaur (Ph.D, M.Phil, B.Ed., M.A., B.Com.).
pages cm
Includes bibliographical references.
ISBN 9789390384044 (Hardbound)

1. Agricultural subsidies--India. I. Title.

HD2073.R35 2016 DDC 338.1854 23

*Published by* : **Scholars World**
*A Division of*
**Astral International Pvt. Ltd.**
– ISO 9001:2008 Certified Company
4736/23, Ansari Road, Darya Ganj
New Delhi-110 002
Ph. 011-43549197, 23278134
E-mail: info@astralint.com
Website: www.astralint.com

*Laser Typesetting* : **Classic Computer Services,** Delhi - 110 035

*Printed at* : **Replika Press Pvt Ltd**

DEDICATED TO

**WAHEGURU**

**AND**

**MY RESPECTED PARENTS**

# Acknowledgement

Successful completion of this book involves interests and efforts of many people, which makes it obligatory on my part to record my thanks to them.

I wish to express my profound gratitude to the Almighty **God** with whose grace and blessing, I have been able to complete another chapter of my life. Words cannot express the deep sense of gratitude and indebtedness.

I am indebted to my parents S. Ram Singh and Mrs. Darshan Kaur for inculcating in me the dedicating and discipline to do whatever I undertake. I am also thankful to my younger brother Er. Manjinder Singh and husband S. Dalvir Singh, who made all efforts to get my book completed.

Last, but not the least the contribution of my friends is also beyond the limits of words and sentences. I extend thanks and appreciation to everyone who have helped me directly or indirectly to get this work done at a given time.

Suggestions for improvement are solicited and would be incorporated in future editions, wherever possible.

***Dr. Rajwinder Kaur***

# Preface

It gives immense pleasure to me in presenting the book entitled *"Analysis of Agricultural Subsidies in India"* for the students who are doing research work related to agricultural subsidies. I am convinced that researchers face a lot of difficulties in understanding the contents of subsidies. Therefore, a humble attempt has been made to bring out a simplified book. Every chapter has been designed and developed according to the need of the researchers. I shall feel amply rewarded if he book is profitably used by all those for whom it is written.

I am highly thankful and obliged to Astral International Pvt. Ltd., New Delhi for publishing the book in time. I will welcome any criticism of the subject matter and style for further improvement.

***Dr. Rajwinder Kaur***

# Contents


| | | |
|---|---|---|
| | *Acknowledgement* | *vii* |
| | *Preface* | *ix* |
| | *List of Tables* | *xiii* |
| 1. | **Introduction** | **1** |
| 2. | **Need of the Study** | **7** |
| 3. | **Gross Cropped Area in India** | **31** |
| 4. | **Agricultural Subsidies in India** | **41** |
| 5. | **Agricultural Subsidies and Productivity in India** | **97** |
| 6. | **Summary, Conclusions and Policy Implications** | **111** |
| | ***Appendices*** | **125** |
| | ***Bibliography*** | **139** |

# List of Tables

Table 3.1: Zone-wise Distribution of Gross Cropped Area in India during 1980-81 to 2006-07 (In 000 hectares) 32

Table 3.2: State-wise Distribution of Gross Cropped Area in South Zone in India during 1980-81 to 2006-07 (In 000 hectares) 33

Table 3.3: State-wise Distribution of Gross Cropped Area in West Zone in India during 1980-81 to 2006-07 (In 000 hectares) 35

Table 3.4: State-wise Distribution of Gross Cropped Area in North Zone in India during 1980-81 to 2006-07 (In 000 hectares)) 36

Table 3.5: State-wise Distribution of Gross Cropped Area in East Zone in India during 1980-81 to 2006-07 (In 000 Hectares) 38

Table 3.6: State-wise Distribution of Gross Cropped Area in North-East Zone in India during 1980-81 to 2006-07 (In 000 Hectares) 39

Table 4.1: Distribution of Total Subsidies in India during 1980-81 to 2008-09 (In Rs. Crores) 42

Table 4.2: Zone-wise Distribution of Total Subsidies in India during 1980-81 to 2008-09 (In Rs. Crores) 43

Table 4.3: State-wise Distribution of Total Subsidies in South Zone in India during 1980-81 to 2008-09 (In Rs. Crores) 44

Table 4.4: State-wise Distribution of Total Subsidies in West Zone in India during 1980-81 to 2008-09 (In Rs. Crores) 45

Table 4.5: State-wise Distribution of Total Subsidies in North Zone in India during 1980-81 to 2008-09 (In Rs. Crores) 46

Table 4.6: State-wise Distribution of Total Subsidies in East Zone in India during 1980-81 to 2008-09 (In Rs. Crores) 47

Table 4.7: State-wise Distribution of Total Subsidies in North-East Zone in India during 1980-81 to 2008-09 (In Rs. Crores) 49

Table 4.8: Zone-wise Distribution of Fertilizers Subsidies in India during 1980-81 to 2008-09 (In Rs. Crores) 51

Table 4.9: State-wise Distribution of Fertilizers Subsidies in South Zone in India during 1980-81 to 2008-09 (In Rs. Crores) 52

Table 4.10: State-wise Distribution of Fertilizers Subsidies in West Zone in India during 1980-81 to 2008-09 (In Rs. Crores) 53

Table 4.11: State-wise Distribution of Fertilizers Subsidies in North Zone in India during 1980-81 to 2008-09 (In Rs. Crores) 55

Table 4.12: State-wise Distribution of Fertilizers Subsidies in East Zone in India during 1980-81 to 2008-09 (In Rs. Crores) 56

Table 4.13: State-wise Distribution of Fertilizers Subsidies in North-East Zone in India during 1980-81 to 2008-09 (In Rs. Crores) 57

Table 4.14: Zone-wise distribution of Fertilizers Subsidies in India during 1980-81 to 2008-09 (In Rs./Hectare) 59

Table 4.15: State-wise distribution of Fertilizers Subsidies in South Zone in India during 1980-81 to 2006-07 (In Rs./Hectare) 60

Table 4.16: State-wise distribution of Fertilizers Subsidies in West Zone in India during 1980-81 to 2006-07 (In Rs./Hectare) 62

Table 4.17: State-wise distribution of Fertilizers Subsidies in North Zone in India during 1980-81 to 2006-07 (In Rs./Hectare) 63

Table 4.18: State-wise distribution of Fertilizers Subsidies in East Zone in India during 1980-81 to 2006-07 (In Rs./Hectare) 64

Table 4.19: State-wise distribution of Fertilizers Subsidies in North-East Zone in India during 1980-81 to 2006-07 (In Rs./Hectare) 65

Table 4.20: Zone-wise Distribution of Electricity Subsidy in India during 1980-18 to 2008-09 (In Rs. Crores) 68

Table 4.21L State-wise Distribution of Electricity Subsidy in South Zone in India during 1980-81 to 2008-09 (In Rs. Crores) 69

Table 4.22: State-wise Distribution of Electricity Subsidy in West Zone in India during 1980-81 to 2008-09 (In Rs. Crores) 70

Table 4.23: State-wise Distribution of Electricity Subsidy in North Zone in India during 1980-81 2008-09 (In Rs. Crores) 71

Table 4.24: State-wise Distribution of Electricity Subsidy in East Zone in India during 1980-81 to 2008-09 (In Rs. Crores) 72

Table 4.25: State-wise Distribution of Electricity Subsidy in North-East Zone in India during 1980-81 to 2008-09 (In Rs. Crores) 73

Table 4.26: Zone-wise Electricity Subsidy in India during 1980-81 to 2000-01 (In Rs./Hectare) 74
Table 4.27: State-wise Electricity Subsidy in South Zone in India during 1980-81 to 2000-01(In Rs./Hectare) 75
Table 4.28: State-wise Electricity Subsidy in West Zone in India during 1980-81 to 2000-01 (In Rs./Hectare) 76
Table 4.29: State-wise Electricity Subsidy in North Zone in India during 1980-81 to 2000-01 (In Rs./Hectare) 77
Table 4.30: State-wise Electricity Subsidy in East Zone in India during 1980-81 to 2000-01 (In Rs./Hectare) 78
Table 4.31: State-wise Electricity Subsidy in North-East Zone in India during 1980-81 to 2000-01 (In Rs./Hectare) 79
Table 4.32: Zone-wise Distribution of Irrigation Subsidy in India during 1980-81 to 2006-07 (In Rs. Crores) 81
Table 4.33: State-wise Distribution of Irrigation Subsidy in South Zone in India during 1980-81 to 2006-07 (In Rs. Crores) 82
Table 4.34: State-wise Distribution of Irrigation Subsidy in West Zone in India during 1980-81 to 2006-07 (In Rs. Crores) 83
Table 4.35: State-wise Distribution of Irrigation Subsidy in North Zone in India during 1980-81 to 2006-07 (In Rs. Crores) 84
Table 4.36: State-wise Distribution of Irrigation Subsidy in East Zone in India during 1980-81 to 2006-07 (In Rs. Crores) 86
Table 4.37: State-wise Distribution of Irrigation Subsidy in North-East Zone in India during 1980-81 to 2006-07 (In Rs. Crores) 87
Table 4.38: Zone-wise Distribution of Irrigation Subsidy in India during 1980-81 to 2006-07 (In Rs./Hectare) 88
Table 4.39: State-wise Distribution of Irrigation Subsidy in South Zone in India during 1980-81 to 2006-07 (In Rs./Hectare) 89
Table 4.40: State-wise Distribution of Irrigation Subsidy in West Zone in India during 1980-81 to 2006-07 (In Rs./Hectare) 90
Table 4.41: State-wise Distribution of Irrigation Subsidy in North Zone in India during 1980-81 to 2006-07 (In Rs./Hectare) 91
Table 4.42: State-wise Distribution of Irrigation Subsidy in East Zone in India during 1980-81 to 2006-07 (In Rs./Hectare) 92
Table 4.43: State-wise Distribution of Irrigation Subsidy in North-East Zone in India during 1980-81 to 2006-07 (In Rs./Hectare) 94
Table 5.1: Zone-wise Distribution of Total Subsidies and Productivity during 1980-81 to 2006-07 (Subsidies (Subs.) in Rs./Hectare, Productivity (Prod.) in Kgs./Hectare) 99

Table 5.2: State-wise Distribution of Total Subsidies and Productivity in South Zone in India during 1980-81 to 2006-07 (Subsidies (Subs.) in Rs./ Hectare, Productivity (Prod.) in Kgs./Hectare) 101

Table 5.3: State-wise Distribution of Total Subsidies and Productivity in West Zone in India during 1980-81 to 2006-07 (Subsidies (Subs.) in Rs./ Hectare, Productivity (Prod.) in Kgs./Hectare) 103

Table 5.4: State-wise Distribution of Total Subsidies and Productivity in North Zone in India during 1980-81 to 2006-07 (Subsidies (Subs.) in Rs./ Hectare, Productivity (Prod.) in Kgs./Hectare) 105

Table 5.5: State-wise Distribution of Total Subsidies and Productivity in East Zone in India during 1980-81 to 2006-07 (Subsidies (Subs.) in Rs./ Hectare, Productivity (Prod.) in Kgs./Hectare) 107

Table 5.6: State-wise Distribution of Total Subsidies and Productivity in North-East Zone in India during 1980-81 to 2006-07 (Subsidies (Subs.) in Rs./Hectare, Productivity (Prod.) in Kgs./Hectare) 109

Table A.1: Availability, Imports and Subsidies of Fertilisers in India during 1980-81 to 2008-09 127

Table A.2: State-wise Distribution of Fertilizers Consumption in South Zone in India during 1980-81 to 2008-09 (In 000 tonnes) 128

Table A.3: State-wise Distribution of Fertilizers Consumption in West Zone in India during 1980-81 to 2008-09 (In 000 tonnes) 129

Table A.4: State-wise Distribution of Fertilizers Consumption in North Zone in India during 1980-81 to 2008-09 (In 000 tonnes) 130

Table A.5: State-wise Distribution of Fertilizers Consumption in East Zone in India during 1980-81 to 2008-09 (In 000 tonnes) 131

Table A.6: State-wise Distribution of Fertilizers Consumption in North-East Zone in India during 1980-81 to 2008-09 (In 000 tonnes) 132

Table A.7: State-wise Distribution of Expenditure and Receipt in South Zone in India during 1980-81 to 2006-07 (In Rs. Crores) 133

Table A.8: State-wise Distribution of Expenditure and Receipt in West Zone in India during 1980-81 to 2006-07 (In Rs. Crores) 134

Table A.9: State-wise Distribution of Expenditure and Receipt in North Zone in India during 1980-81 to 2006-07 (In Rs. Crores) 135

Table A.10: State-wise Distribution of Expenditure and Receipt in East Zone in India during 1980-81 to 2006-07 (In Rs. Crores) 136

Table A.11: State-wise Distribution of Expenditure and Receipt in North-East Zone in India during 1980-81 to 2006-07 (In Rs. Crores) 137

# Chapter 1
# Introduction

The socio-economic structure, which prevailed prior to the British rule in the country, resulted in the organization of self-sufficient villages. It has been maintaining some kind of static equilibrium. The Indian peasant, though not properly educated, has adequate experience of farming systems and he has been dependent on it for the means of living. The Royal Commission of Agriculture in India observed that both the methods of cultivation and social organization exhibit that settled order which is characteristic of all countries in which the cultivating peasant has long lived in and closely adapted himself to the conditions of a particular environment.

The British government in India did not bother for the development of agriculture in this country. The East India Company had also done nothing it was primarily interested in trading and exploitation of resources of the country. The main objective of the British policy was administrative consolidation other than economic regeneration till the end of the nineteenth country, no attention was paid to agriculture (Mishra, 2006).

The Indian agrarian economy on the eve of independence was critical in situation. It could be characterized totally primitive, deteriorative and turbulent. During the British imperial regime, no pervasive and conductive measures were taken to boost the agriculture. At the time of independence, Indian economy was in the worst state of affairs, the deficiency of food grains was quite alarming and aggravating (Chahal, 1999).

The partition of country worsened the food situation in the country. This reduced the agricultural production and created difficulties both for food grains and commercial crops. The country was left with 82 per cent of the total population of undivided India as well as only with 69 per cent of land under rice, 65 per cent under wheat and 75 per cent under all cereals. The cultivators were under heavy debt and most of the holdings were uneconomic (Chahal, 1999).

Agricultural development is a condition precedent for the overall development of the economy. A progressive agriculture serves as a powerful engine of economic growth. It helps in initiating and sustaining the development of other sectors of the economy by providing necessary capital, labour, raw material, wage goods and foreign exchange (Kumar, 2007).

In view of this, after independence tremendous efforts were made to boost the economy through agriculture as one of the tools for development. The Government of India adopted a positive approach and hence a well defined policy of integrated production programmes with defined targets and a proper distribution programme was adopted along with other measures for the overall economic development of the country. Specific programmes like new agriculture technology were introduced to convert agriculture into a successful and prosperous business, to bring more land under cultivation and to raise agriculture production (Singh, 1994).

In India, the adoption of new agricultural technique was costly than that of traditional method of cultivation. In traditional method, inputs were least expensive, on the other hand, inputs in modern technology like high yielding varieties of seeds, fertilizers, farm mechanization and irrigation were very costly and Indian farmers being poor were not in a position to buy these expensive inputs. On the recommendations of food grain price committee (Jha Committee), the Government of India started the scheme of subsidies on purchase of various agriculture inputs to facilitate the farmers (Singh, 1994).

Subsidies, by means of creating a wedge between consumer prices and producer costs, lead to changes in demand/supply decisions. Subsidies are often aimed at inducing higher consumption/production and offsetting market imperfections including internalization of externalities, achievement of social policy objectives including redistribution of income (Gulati, 2007).

Subsidies on farm inputs were initially meant to induce farmers to adopt new technology. It was this intention that since the mid-1970s, central and states governments have followed a policy of supplying fertilizers, power, irrigation and credit at prices which do not fully cover costs. No doubt, input subsidies have helped in large scale adoption of new technology and output growth. However, their levels have risen to proportions which cannot be sustained and their beneficial effects are said to be overweighed by the adverse effect in terms of macro-economic balances, slowing down of public investments in agriculture, inefficient use of resources, degradation of environment and reduction of employment (Rao, 1994).

Subsidies have occupied agricultural economists for a long time because they are pervasive in agriculture, even though they are often applied in ways that benefit mostly richer farmers, cause inefficiencies, lead to a heavy fiscal burden, distort trade, and have negative environmental effects. Agricultural subsidies can play an important role in early phases of agricultural development by addressing market failures and promoting new technologies (Fan, 2008).

All of these subsidies by reducing the prices of the inputs, served in the initial stages of green revolution, as incentives to the farmers for adopting the newly introduced seed-cum-fertilizer technology. These helped in raising the agricultural output, after some time, the amount paid on these subsidies began to rise (Gulati, 2003).

Viewed in terms of pure domestic economy, the input subsidies have often been accused of causing most harmful effect in terms of reduced public investment in agriculture on account of the erosion of investible resources, and wasteful use of scarce resources like water and power. Further, apart from causing unsustainable fiscal deficits, these subsidies by encouraging the intensive use of inputs in limited pockets have led to lowering the productivity of inputs, reducing employment elasticity of output through the substitution of capital for labour and environmental degradation such as water logging and salinity, on the other hand, and lowering of water tables (Gulati, 1995).

In India, at present, centre as well as state governments are providing subsidies on fertilizers, irrigation (canal water), electricity and other subsidies to marginal farmers and farmers' cooperative societies in the form of seeds, development of oil seeds, pulses, cotton, rice, maize and crop insurance schemes and price support schemes etc. Out of these subsidies, the central Government of India provides indirect subsidies to farmers on the purchase of fertilizers from 1977, whereas state governments are providing subsidies on irrigation as well as on electricity (Government of Punjab, Agriculture Department, Chandigarh).

## Agricultural Subsidies by Central Government

At present, the central government pays subsidies to the farmers on the purchase of fertilizers. Fertilizers are an important component of agricultural technology. Whereas initially organic fertilizers were mainly used in the fields, however, chemical fertilizers have played a very important role in enhancing the agricultural production. To ensure availability of fertilizers to farmers at affordable prices, the government of India provides huge subsidies to fertilizers manufacturing industry from 1977. Hence, the government of India provides indirect subsidies to farmers for the purchase of fertilizers (State Environment of Punjab – 2005).

## Agricultural Subsidies by State Governments

Many years ago, state governments were considered to be only police states. Their function was mainly to maintain law and order in the country and to protect it from foreign invasion. So far as the economic affairs of the country were concerned, a complete laissez faire policy was followed. However, the situation has changed now. The concept of welfare state has taken deep roots. The central government encouraged the strategy of enhancing food grains production in states, particularly wheat and rice, for meeting the emergent food demand in the country. Punjab state leads other states in terms of contribution of wheat and rice to central pool (Karnik, 1996).

For the development of agricultural sector, at present, Punjab Government is giving subsidy on electricity as well as on irrigation. Energy in the form of electricity plays a key role in performance of agricultural sector, in Punjab as it is used in pumping ground water for irrigation purposes. At present government of Punjab is giving electricity to Punjab farmers, free of cost, through Punjab state electricity board, now is unbundled into Punjab state power cooperation limited and in Punjab state transmission cooperation limited (Government of Punjab, Punjab State Electricity Board). The water being supplied to the farmers for irrigation purpose is free of cost. It is quite difficult to estimate the values of this water being supplied to agriculture sector, therefore, total amount of subsidy on this account could not worked out on actual basis (Government of India, Pricing Water Policy, 2010).

## Agricultural Subsidies by Agriculture Department

At state level, agriculture departments are also giving subsidies on various inputs *i.e.* gypsum, seeds, gobar gas plants, farm mechanisations, pesticides under various schemes as facilities to the farmers. At present, Agriculture Department of Punjab, Chandigarh is providing these subsidies to Punjab farmers.

The use of gypsum has been wide spread in practice and farmers are trying to convert their waste lands into productive lands. The subsidy of Rs. 1,000 per hectare or 50 per cent of total cost of gypsum is provided. Jantar seeds are provided with free of cost, 100 per cent subsidy on oil seeds and 50 per cent subsidy on the cost of hybrid seeds of pulses and wheat. On gobar gas plant Rs. 2,610, per plant are given as subsidy (Government of Punjab, Agriculture Department, Chandigarh).

Farm mechanisation has been helpful to bring about a significant improvement in agricultural productivity. Thus there is a strong need for mechanisation is useful for farm operations like harvesting, threshing etc. A scheme "Promotion of Agricultural Mechanization among small size category farmers" has been launched during 1992-93. Under this scheme a subsidy of 30 per cent subject to limit of Rs. 30,000 is given for the purchase of tractors up to 30 powers take off horse power. The subsidy of Rs. 30,000 is provided on rotavetors and Rs. 15,000 on seed drill machine (Government of Punjab, Agriculture Department, Chandigarh).

The total agriculture subsidies have increased from Rs. 14,069 crores in 1993-94 to Rs.17,677 crores in 1995-96 and further increased to Rs.1,60,917 crores in 2008-09. The centre government's deficit has increased from Rs.8,299 crores in 1980-81 to Rs.21,858 crores in 1985-86 and further increased to Rs.1,18,816 crores in 2000-01. In 2011-12, it will be Rs.412,817 crores (Budget Estimate, 2011), whereas state's deficit has increased from Rs.3.713 crores in 1980-81 to Rs.7,521 crores in 1985-86 and further increased to Rs.90,098 crores in 1999-2000, it declined to Rs.87,922 crores in 2000-01, to Rs.1,20,631 crores in 2003-04 and further increased to Rs.1,99,510 crores in 2009-10 (Reserve Bank of India, 2011).

Fiscal deficit in India has to be reduced, for reducing the fiscal deficit, the Government of India has to reduce its expenditure and increase the revenue. So far as the reduction in expenditure for the agriculture sector is concerned, it can be brought out mainly by reduction in subsidies. The new economic policy 1991, was the first policy which recommended that in order to reduce fiscal deficit of Government, efforts were made to reduce the agriculture subsidies. Again in May 1997, the Finance Minister Mr. P. Chidambaram issued a 'white paper' entitled 'Government Subsidies in India'. This paper stated subsidies as the cause of inefficient and wasteful use of resources. It listed the subsidies granted for fertilizes, irrigation and power as subsidies on non-merit economic services and stated that the main purchase of these subsidies is to help the farmers to reduce the production cost, but the fact is that big farmers are getting more benefits than that of small size category farmers from these subsidies, who already have the capacity to pay for purchase of inputs, then these subsidies should be abolished in a phased manner (Aggarwal, 2007).

After these recommendations neither central Government nor State Governments reduce or withdraw these subsidies. In fact the disappointing piece of the whole story is that agriculture subsidies are not due to economic necessity but because of political expediency (Gulati, 2003).

## Chapter 2

# Need of the Study

Subsidies are often criticized for their financial burden. Some researchers assert to the extent that these should be withdrawn in a phased manner, such a step will reduce the fiscal deficit, improve the efficiency of resources use, funds for public investment in agriculture. On the other hand, there is a fear that agriculture production and income of farmers would decline if subsidies are curtailed. These are very important issues, which need serious investigation.

Before analysing the State-wise agriculture subsidies in India, I would like to discuss the research done in the past of various researchers in the same field.

Beacuse past theory and practice is necessary when conducting any research work. It provides information of the work done in the related area and the theoretical frame work on which the proposed solution of the problem can be based. The relevant literature was reviewed in detail to understand the nature and extent of the work done on the related topic. An attempt is made to analyse the nature of the work done during past in the related field. The brief review of literature has been given as under:

Yadav, (1982) determined the impact of Agricultural subsidies on the incomes of small and marginal farmers in Ajitwal block in Etawah district of Uttar Pradesh in 1980. Data was collected from 30 beneficiary and 30 non-beneficiary small and marginal farmers from five villages. Data was collected through randomly. The author compared the beneficiary and non-beneficiary farmers and found due to subsidies the income of farmers increased. This study showed that income beneficiary farmers were about 70 per cent more than of non-beneficiary farmers. At the end, the author suggested the provision of subsidy for small and marginal farmers only.

Sharma, (1982) examined the impact of agricultural subsidies on national income and agricultural production. For this purpose the author used the time period from 1970-71 to 1981-82 and a general equilibrium model. The study revealed that during this period, agricultural subsidies affected the national income and agriculture production positively. The author estimated that the co-efficient of fertilizer subsidy was not statistically significant, even at low level of probability. The author suggested that money to different subsidies should be allocated according to the productivity of various subsidies and in developing country, there may be a possibility of misuse of agricultural subsidies and therefore, their continuance finally led to inflationary pressure in the economy.

Raj, (1984) stated that owing to large investment with long gestation period, self-sufficiency in food-grains cannot be achieved in the short-term period through improvement in physical and institutional infrastructure. Under such conditions government always prefers to adopt short-term policies, such as support product prices policy and subsidizing inputs with a view to introducing the farmers to increase food-grains output along with existing production function. In this study the authors made an attempt to illustrate effects of such policies by analysing the programmes of price support and fertilizer subsidy for achieving self-sufficiency in wheat in the country. The authors compared the price support and input subsidy approaches to achieve self-sufficiency in wheat production by using the criterion of (i) total cost of the government (ii) total social benefit and cost of the programmes (iii) distribution of benefits and (iv) foreign exchange savings. This study concluded that input subsidy programme was found to be more appropriate considering total cost of the government, overall social benefit and cost of the programmes and distribution of benefits, on the other hand price support programme was found more feasible than input subsidy taking into consideration the foreign exchange saving criterion.

Gupta, (1984) tried to analyse the agricultural subsidies in India from 1970-71 to 1982-83. The author used linear regression model. The study showed that during this period, the use of agricultural subsidies increased at faster rate but there was a large inter-state disparity. It was found that Punjab, Uttar Pradesh and Maharashtra used about half of the total agricultural subsidies but accounted for only 30 per cent of the gross cropped area of the nation. Whereas Rajasthan, Madhya Pradesh and Odisha received only nine per cent of agricultural subsidies but claimed 27 per cent of the gross cropped area of the nation. The study also showed that there was inter-state disparity in the use of agricultural subsidies per hectare of gross cropped area. It was found to be the highest in Punjab (Rs. 216.18) and lowest in Rajasthan (Rs. 12.45). The author observed that the benefits of fertilizers subsidies were found to be biased against the small and marginal farmers and the author suggested that more agricultural subsidies should be given to poor states and small and marginal farmers to encourage them to utilize more inputs at lower costs.

Sirohi, (1984) stated that the social justification of the subsidies is only when they promote agricultural development and bring about equitable distribution of income. As regards the effects of various input subsidies on agricultural production and national income, the largest favourable impact was observed in case of subsidy on electricity. The equity effects of the input subsidy seem to have gone mainly to medium and big farmers. While fertilizer subsidy aggravated the income disparity in the rural areas, the benefits of irrigation accrued to all classes except the big farmers whose income declined. It thus brought equity, on the other hand aggravated regional disparity because of the benefits of fertilizer and irrigation accrued to regions with better irrigation facilities.

Sidhu, (1985) tried to examine the relative merits of price support versus fertilizer subsidy policy for food self-sufficiency in India. Rice and wheat crops are selected because of their high productivity and area as compared to other crops. These crops are covered under price support programme and use about 65 per cent of the total fertilizers. This study provided decision criteria for policy makers to select between the price-support and fertilizers subsidy under different situations. Fertilizer subsidy policy is found to be better price support policy in terms of net social benefits and benefit-cost ratio. But from the view point of the foreign exchange savings, price-support policy occupies better plashes. Fertilizer subsidy policy again turns out to be better as compared to price support policy due to low price elasticity of output supply, higher price elasticity of fertilizer. A more egalitarian distribution of procedures income is deserved under fertilizer subsidy policy, it also turns out to be anti-inflationary.

Deininger, (1986) examined the slowdown in agricultural growth and the macroeconomic impact if rising fiscal deficits has refocused attention to public expenditures in the agricultural sector. Rising levels of agricultural subsidies has been blamed for crowding out much needed productivity enhancing investments. The study examined the potential welfare impacts of subsidy reforms by tracing the beneficiaries, (the farmers and consumers) of food grain price subsidies and by assessing the distribution and level of these subsidies across households at the state level. Using benefit incidence analysis, author found that agricultural subsidies benefited only a few states and larger farmers within these states. The author suggested that agricultural subsidies should be given to poor states and to small and marginal farmers.

Gulati, (1989) made an attempt to estimate the quantum and distribution of input or agricultural subsidies across states in India during 1980's. The dispersion pattern of input subsidies has implications for incentive structures prevailing in the agricultural sectors of different states as well as efficiency in production of agricultural commodities across regions and states. This study covered three major inputs of modern agriculture: fertilizers, irrigation and electricity. The concept of subsidy on these inputs was defined in a more economically meaningful sense, which differs significantly from one generally delineated in government budgets. The author examined and gave his findings which can be significant in evolving a rational describable cropping pattern in different agro climate zones based on the principle of comparative advantage. The study revealed that total input subsidy, averaged over seven years, 1980-81 to 1986-87, turns out to be about Rs.9,000/- crores at all India level. It is approximately 17 per cent of net value added in Indian agriculture. More than 70 per cent of total input subsidy is on irrigation through major and medium scheme. The author analysed that there is unequal distribution of agricultural subsidies as well as in GCA. The author found that at state level, the percentage share of Uttar Pradesh, Andhra Pradesh and Punjab in total subsidy comes to about one-third while they account for only one-fourth of all India gross cropped area. Input subsidies as a percentage of State domestic product in agriculture averaged over 1980-81 to 1986-87, are highest for Tamil Nadu (31.7 per cent), followed by Punjab (24.5 per cent), Haryana (23 per cent), Andhra Pradesh (21.3 Per cent) and Uttar Pradesh (18.2 per cent). At the bottom end are the states like Himachal Pradesh (2.0 per cent) Assam (2.4 per cent) and Jammu & Kashmir (5.4 per cent).

Sharma, (1990) revealed in this study that subsidies have become unsustainable. In order to release resources for higher investments in the agricultural sector, large scale price and institutional reforms are needed to relieve the pressure of subsidies on the exchequer. Under the circumstances, it makes much sense to improve terms and trade for agriculture and complement this by stepping up investment in agriculture through reduction in subsidies. The increased investment in agriculture appears to be a better bargain than short-sighted measures such as subsidies. This is because of the fact that cultivable land in India is in short supply and raising productivity per unit of cultivable area will require heavy investments in irrigation, rural infrastructure research and extension. The author further analysed that investments in basic infrastructure correct for regional imbalances and promote greater equity at farm level, while subsidies tend to accentuate inequality.

Mukherji, (1990) tried to examine economic of electricity subsidy in West Bengal. He has conducted a study to describe the electricity subsidy in West Bengal. The author used primary data from 40 villages and he has selected 580 respondents from those villages. The author found that electricity subsidies benefit only big farmers than that of the small size category farmers. The author suggested that electricity subsidy should be given to small size category farmers only.

Sagar, (1991) argued that the agricultural sector in the developed countries has enjoyed a greater degree of production than the import competing manufacturing sectors. Usually this is attributed to string farm lobbies and hence on political factors. They provided a theoretical model and a possible explanation of this phenomenon based on purely economic arguments. Two importable are accommodated in a three-good three-factor model of trade and production *i.e.* one is a Labour intensive manufacturing good and other is an agricultural commodity. This captures the trade pattern of a typical industrialized country with an agricultural sector such as Europe and the United State America. They showed that uniform tariffs on subsidies in agriculture and labour intensive manufacturing will definitely hurt the land owners in real terms and may reduce their absolute return. The author suggested that, if there has to be protection, it should be biased in favour of agriculture.

Malik, (1993) stated that the conventional agricultural system in the semiarid administrative district of Ludhiana in Punjab, depended on heavy doses of inorganic fertilizers and pesticides, repeated deep ploughing and heavy use of ground water. Electricity subsidies promoted excessive water use (15 per cent more than recommended levels). Over-irrigation and large-scale rice production were decreasing the groundwater tables by 0.8 m/year. This case study analysed alternatives to the conventional paddy-wheat production system. The alternatives were 18 combinations of tillage, irrigation and fertilization practices. As for water use, the researchers compared overuse, recommended use, and less than recommended use. They also compared financial and economic values for each farming practice. The five policy options tested were current policy, removing commodity support for consumers, removing input subsidies, removing commodity and input subsidies and free world trade. The conventional practice under the current policy lose 25 per cent of its income to groundwater depletion, soil degradation and off-side environmental costs, while the income from the most resources–conserving practice. (reduce tillage, inorganic fertilizer plus farm yard manure, 80 per cent of recommended use) has a 9 per cent net farm income and has no environmental costs. Further analysis showed that net income falls the most for the predominant practice which relies on government subsidies. The net present value of groundwater depletion estimated at subsidized electricity rates equals around six per cent of gross operating margin. Thus, the author concluded that pricing reform, research on water management and alternative cropping systems, and improved monitoring of groundwater levels were required to achieve sustainable agricultural production in Ludhiana.

Majumdar, (1993) observed the growing burden plashed on Indian Public finances by the subsidization of fertilizer has become one of the most concerning aspects of the Indian Economy. In this study, the author utilized a component type analysis to determine the percentage share of subsidy benefiting both the farm and non-farm sectors. He also looked at the major factors behind the increase in subsidy, the scope for reducing the subsidy through the feed-stock mechanism and considered the extent to which the burden created by subsidies can be switched to the farmer. Under certain assumptions the author found that the subsidy component of farm and non-farm sectors is in the ratio of 2:1. The component of subsidy going to the non-farm sector is mainly accounted for by the feed stock sector rather than the industrial sector. He suggested for policy that eliminates the fertilizer subsidy that neither pampers the farm sectors on the feed-stock sector and that this should evolve from discussions involving representative organizations of those affected academics and experts.

Kalra, (1993) conducted a study to analyse the growth of fertilizer subsidy in India has come to plashes on intolerable strain on central Government finances and has raised questions concerning its sustainability in the long term. The study considered the position and effectiveness of Indian fertilizer subsidies both pertaining to equity and efficiency. Fertilizer subsidies in government accounts were shown to be the actual price received by the industry minus the price level dictated purely by efficiency namely the imports price level. This study showed the farmer has in net terms been toned rather than subsidized under the assumption of the free trade scenario. This study concluded that encouraging the expansion of irrigation facilities was a more cost effective and efficient measure in terms of raising the level of agricultural of production. The author suggested that the subsidy as burden on government can be mitigated by improving the efficiency of the fertilized Industry, as well as efficiency of distribution and consumption of fertilizer.

Murgai, (1995) stated that most of India's agricultural subsidies are both inefficient and regressive while the power subsidy to agriculture has been increasing over time, the other input subsidies on fertilizer and irrigation increased over the eighties but has been reduced in the nineties. The fertilizers subsidy increased from

0.3 per cent of gross domestic product in 1981-82 to 1.1 per cent in 1989-90, corrective measures were taken in response to the balance of payments crisis in the early nineties and there was a crores pending reduction in the early years of the nineties in the fertilizer subsidy bill to about 0.7 per cent of gross domestic product. The irrigation subsidy increased from 0.3 per cent in 1980-81 to 0.4 per cent in 1990-91 and by 1999-2000 has fallen back to 0.3 per cent, likely because of the slowdown in irrigation investments. In this study, the author suggested that agricultural subsidies should be replaced by better services and more investments.

Sant, (1996) described that the power sector reforms regarding tariff of electricity or electricity subsidy. An analysis of irrigation pump sets subsidy users in Maharashtra State was done. During analysis, it was observed that with efficiency improvements, most of the farmers will be able to pay the electricity charges. It was also observed that most of the farmers can pay the costs of electricity much higher that usually believed. Therefore, the opposition by state Electricity Board's need to be seen as opposition to making the State Electricity Board's accountable for the electricity lower and theft.

Schwartz, (1997) examined the problems of defining and measuring government subsidies and also examined why and how government subsidies are used as a fiscal policy tool. The author discussed their general economic effects in terms of real welfare costs and distribution implications, appraised international empirical evidence on government subsidies and offered options for their reform. Recent international trends in government subsidy expenditure are analysed for the 16 years period from 1975 to 1990, using general government subsidy data for 60 countries from the united nation's System of National Accounts (SNA). They retrieved major policy options for subsidy reform, focusing on ways to improve the cost-effectiveness of subsidy programme.

Kaushik, (1998) pointed out that agricultural subsidies are a worldwide phenomenon, a creation of modern economies, which make stand - alone agriculture the least remunerative occupation. The developed countries subsidies their agriculture on a much larger scale than India does. In 1999, subsidy payments to agriculture were $ 54.0 billion in the U.S., $ 114.5 billion in EU countries and $ 58.9 billion in Japan, against $ 7.2 billion in India. On a per farmer basis, the subsidy was $ 21,000 in US, $ 17,000 in EU, $ 26,000 in Japan and measly $ 66 in India. Expressed as percentage of the value of agricultural production, the subsidy was 24 per cent in US, 49 per cent in EU and 65 per cent in Japan, but just 6.5 per cent in India. The Government controls the minimum retail price of fertilizers at a level affordable to the farmers. The author analysed that bulk of the increase in subsidy is in the nature of an intra-economy transfer and to that extent, it is not even a net burden on the exchequer. In this study, the author analysed that small and marginal farmers need to be given special relief by giving them employment in off-farm activities. The income generated from these activities will provide them the much needed cushion for absorbing the unavoidable escalation in input costs. For several years now, an uncertain policy environment has dogged the fertilizer industry in India. This has affected fresh investment and impaired the ability of existing units to plan for the future and take business decision. The removal of price controls will eliminate this uncertainty and pave the way for a stable policy environment. The author suggested that Government may have to continue with agricultural subsidies to pursue the broader goals of food security and protecting the income of farmers, these should be given to them directly, the only way to bring peace to this vital industry and enable it to serve Indian agriculture effectively.

Mehta, (1998) tried to examine that irrigation and power being state subjects, the analysis was carried out for three states Gujarat, Punjab and Uttar Pradesh. The choice of the states was based on the fact that, all the three has large agriculture sectors, Moreover, the three states were districts in terms of their sources of irrigation and the power pricing policies. The author found that rapid rural electrification along with subsidized power and irrigation for agriculture has led to high dependence on underground water sources. The overuse of groundwater especially in some districts, was merging as a grave concern. The author found that during 1985, out of 183 total blocks in Gujarat, 6 blocks were over exploited which increased to 45 blocks in 1994. In Punjab, out of 118 total blocks, 64 blocks were over exploited, during 1985 which incurred its 73 blocks in 1994. Similarly in Uttar Pradesh, out of total 895 blocks, 53 blocks were over exploited in 1985 which increased to 65 in 1994. The author showed that moreover highly subsidized or free supply of inputs like power and irrigation has also led to inappropriate cropping patterns. For example in water scarce areas, water intensive sugar cane and rice crops were being promoted. The author suggested that a policy package that is economically and environmentally sound is needed, besides being socially acceptable.

Nayak, (1999) made an attempt to describe the amount of total input subsidies was consistently increasing, both in the case of Punjab and at the national level. From an estimate amount of subsidy of Rs. 504 crores in 1985-86 it has increased to Rs. 1,812 crores in 1995-96 in Punjab. The percentage share of subsidy in agricultural gross domestic product was consistently higher in the case of Punjab than the national average although the gap has narrowed down in the recent years. Still the burden of input subsidies was quite substantial amounting to as much as 11 per cent of agricultural gross domestic product in Punjab. The mounting burden of subsidies in the form of free electricity and free irrigation was taking a severe fall on the state government exchequer. The author concluded that the state expenditures and infrastructure has to take a back seat due to the rise of law and order problems in the region as well as the arbitrary pricing and subsidy policy of the state government.

Gupta, (2000) stated that any reduction in the use of steel cement, automobiles etc. for a couple of years can be taken in the stride, but the substantial decline in the consumption of the fertilizers will has the effect on the food grain production. This was quite clear from the experience of the two consecutive drought years in the 1960s – exploitation in the International market, humiliation suffered at the hands of other countries and handling the large imports. In the year 1996-97, food grain production in India was 200 mn tones and the fertilizer use was 10.7 mn tones for the same production of 200 mn tones of food without the use of fertilizers, 892 mn tones of organic feed has been required, which was virtually impossible. In the absence of fertilizer use food grain production will be lower and import of the food grain from the other countries would be required, this in turn has the devastating effect on balance of payment of country. Now country has achieved self-sufficiency in food grain production supported by increasing consumption of fertilizers. The author suggested that it cannot afford to be complacent and treat fertilizers like any other commodity. Thus there is an urgent need to promote increasing fertilizer use at an accelerated rate in order to maintain self-sufficiency and simultaneous increase in food grain production.

Pal, (2001) tried to analyse the present state of World Trade Organization negotiations and how effective the current World Trade Organization provisions will reduce domestic subsidies in developed countries. He analysed that the broad frame work of subsidy reduction, as outlined in the July package, can be considered as a step towards the right direction but it does not guarantee significant reduction in subsidies. A simulation using the subsidy reduction formulas mentioned in the draft ministerial tent for the Hong Kong meet also indicates that due to the existence of significant overhang between actual and committed levels of subsidies in developed countries, the effective rate of reduction of subsidies will be much less than it appears at the first glance. Unless deep reduction commitments are imposed on developed countries, it will not lead to a substantial cut in their trade distorting domestic supports.

Jogi, (2001) stated that the provision of electricity and irrigation at concessions has encouraged inefficient use of a scarce resources such as water, distorted the inter-temporal resource allocation and promoted spatial, inter-personal and inter-temporal inequities. In Punjab 52.17 per cent of the total blocks in the state were over-exploited and 7.97 per cent of all blocks are dark areas as on 31.3.98. The over-exploitation of underground water has caused a fall in the water table in large parts of the state and this has entailed increased expenditure on deepening of tube wells. In case of canal irrigation it is found that 44 per cent of the water entering the canal has got lost in the canal itself, 27 per cent of the water is wasted by the farmers through excessive use and only 29 per cent is actually used by the crops. The author suggested that if government imposed charges on power and canal irrigation, then the farmers would use natural resources more efficiently. At the same time with the reduction subsidies the government should be able to increase investment in the power sector to improve quality and quantity supplied as well as to increase their efficiency reducing transmission and distribution losses and improving the quality of the service.

Dubash, *et al.* (2001) examined that water related subsidies in agriculture are virtually a universal phenomenon. In this study, the author calculated the per hectare irrigation subsidies in five states of India: Rajasthan, Maharashtra, Andhra Pradesh, Karnataka and Uttar Pradesh. For collecting data, State budget data and Indian National Sample Survey were used. The author analysed that in practice, water tariffs has been set at very low rates. The capital expenditure on irrigation at national level has increased from Rs. 7.65 billion in 1985 to Rs. 110 billion in 2000. In 1997-98, in Maharashtra, irrigation subsidies were Rs. 3108 million, in Uttar Pradesh Rs. 2,777 million, in Andhra Pradesh Rs. 2021 million, in Karnataka Rs. 259 million, in Rajasthan these were Rs. 182 million. The author also found that the magnitude of canal irrigation subsidies per hectare was Rs. 10149 in Maharashtra,

Rs. 1,117 in Uttar Pradesh, Rs. 1387 in Andhra Pradesh, Rs. 337 in Rajasthan and Rs. 242 in Karnataka. The author further analysed that benefits of the subsidies were not distributed equitably, large size category farmers were receiving more subsidy than small and marginal farmers. The author suggested that broad policy and institutional reforms are needed to address the consequences of water related subsidies and a better understanding of the nature of water related subsidies, their magnitude, environmental impact and associated issues of equity is needed.

Kumar, (2002) described that India's food security policy, which has the objective of ensuring food grain availability at an affordable price, has been another compelling reason to provide subsidies for the agricultural inputs, namely, water for irrigation supplied from public systems, electricity used for groundwater pumping and fertilizers. This is because the farmers are denied the option of fetching high prices for their grain through free trading due to the government restrictions on inter-state grain trading and that the only way to make food production remunerative was through cutting down on the input costs. The provision of input subsidies as such is not a bad idea in the agricultural sector. Several countries around the world still report to subsidies to enable farmers to grow food at low cost. To the extent that large subsidies can alter the potential efficiency patterns of water use, water subsidies can cause long term irreversible effects – environmental, physical, geographical and economical. The author suggested that the policies and programmes need to be designed and operationalized to encourage sustainable farming practices with increased use of bio-fertilizers. Subsidies can be introduced for small and marginal farmers to adopt bio-gas plants and scientific compost making community based programmes for increased production of bio-mass from common property wastelands can also be introduced. Extension activities on organic farming practices, bio-gas and low-cost water saving technologies need to be taken up. Agricultural research should shift from a supply driven system to one that takes into account the demand side variables such as the local physical and socio-economic conditions, in order to increase the scope of the search.

Howes, (2002) tried to examine the distribution pattern of electricity subsidy. He has conducted a study to show the distribution pattern of electricity subsidy in farmers of Karnataka State. In this study the author found that per unit of the electricity, subsidy benefited poor farmers the rest went to large size category farmers. Similarly, only two per cent of the subsidy benefited scheduled caste or tribal families. The study found that on average, large size category farmers who own irrigation pump sets received Rs. 29,000 subsidy per year and pump set over as with marginal land-holding received subsidy of approximately Rs. 3,000 per year. The author concluded that electricity subsidies are regressive because large size category farmers are much more likely to have pump sets than small size category farmers and because large size category farmers with pumps use more electricity than small size category farmers with pumps. The author suggested that electricity subsidy should be given to only small size category farmers.

Gupta, (2003) argued that there has been a substantial increase in fertilizer consumption in Punjab in the last three decades Total NPK consumption has increased more than seven fold between 1970-71 and 2005-06 per hectare fertilizer consumption increased from about 37 Kg to 221.7 Kg. in the same period. The fertilizer consumption in Punjab is the highest in India. In addition overuse of nitrogen fertilizer due to higher amounts of subsidy has led to imbalanced use of fertilizes in the state. The $N:P_2O_2:K_2O$ ratio in Punjab is one of the most distorted at 27.8:7.3:1 as against the generally recommended 4:2:1 ratio. The author concluded that intensive use of inputs mainly pesticides and chemical fertilizers which was central to green revolution has created an ecological crisis in the state. If urgent remedial action is not taken, Punjab's agrarian crisis is bound to deepen.

Bhargava, (2003) revealed that just three items of public spending on agriculture – central fertilizer subsidy, electricity subsidy and irrigation subsidy account for nearly one –fourth of the increase in India's public sector deficit in recent years. The contribution of the agricultural sector to recent increase in India's public sector deficit, if other such items like crop insurance, losses, credit etc. are taken into account, will turn out to be for more daunting. According do this study, a vicious circle seems to have already set in. The policy matters has justified agricultural subsidies on the ground of the country's poor's inability to pay market prices for food. But subsidies, by promoting inefficiencies in the use of inputs, has the effect of raising input intensity of farm output and thereby of raising the costs of food production. This in turn, has led to demands for more fiscal favour. What is more, the regime of subsidies has also encouraged rent-seekers to get what they can from the system, with politicians seeking high profile new projects, not proper maintenance of existing systems and farmers employing political pressure to get what give always they can rather than organizing for improved agricultural extension services. In this has contributed to rapidly rising fiscal outgoes. The author suggested that the vicious circle urgently needs to be broken, otherwise it will keep pushing India inexorably towards a situation of rising public sector deficits a situation which may pace the following major risks or a combination of them, for the Indian economy during the 1990s, substantially highest real interest rates, crowding out of some investment, lower growth rates, debt trap substantially higher inflation rates and excessive external debt service burden.

Orden, (2003) stated that since the early 1990s, India has undergone substantial economic policy reforms and economic growth. In this study, author evaluated the protection and support versus dis-protection of agriculture in India. His methodology involved examining market price support for eleven crops, the expenditure on input subsidies benefiting farmers (for fertilizer, electricity and irrigation) and product specific and total producer support estimates over the period 1985-2002. The author drew extensive price-comparison and subsidy measurement data sets. Overall, results indicate that support for agriculture in India has been counter cyclical. Support for agriculture has been rising when world prices are low (as in the mid-1980s and 1998-2002) and falling when world prices are high (as in the early and mid-1990s) Results demonstrate the increased importance of budgetary payments for input subsidies in agriculture in recent years. Yet, in the aggregate for both price support and budgetary expenditures over the period 1985-2002 in the counter-cyclical dimension of agricultural policy dominates a clear trend of movement from dis-protection towards protection. The magnitudes of estimated support for agriculture obtained in this study are important for several reasons. The estimates confirm that high levels of subsidies were required for India to export wheat or rice in recent years.

Sengupta, (2004) stated that India is the third largest producer and consumer of fertilizer in the world. The world fertilizer consumption is approximately 142 million tones and growing at 2 per cent per annum. International trade in fertilizer accounts for approximately 62 million tonnes out of current world consumption of 142 million tonnes *i.e.* approximately 43 per cent of the total consumption. In order to provide fertilizers at an affordable price to farmers, the major thrust of Indian Fertilizer Policy is based on subsidy. The author found that the developed states has got a higher proportion of fertilizer subsidy, which constituted about 60 per cent of total fertilizer subsidy and the similar trend in case of irrigation and electricity subsidy. This study showed that on an average the total input subsidies constituted about 16 per cent of gross domestic product. The author suggested that to sustain the present pace of growth of agricultural production, there is a need of continue the input subsidies by controlling leakages.

Bala, *et al.* (2004) in their study made an attempt to investigate the trends in production and consumption of fertilizers in India and examined the effects of various factors such as price, area under high yielding varieties, gross cropped area, cross irrigated area and subsidy on its consumption. The time series data from 1975-76 to 1999-2000 was taken into account for this study. Results indicated that consumption of fertilizers increased at the rate of 11 per cent over the study period. Whereas the growth rate for fertilizer production was 10.6 per cent. Among the factors affecting fertilizer consumption, subsidy emerged to be the most important factor followed by area under high yielding varieties and gross irrigated area. The author suggested that farmers should be encouraged for balanced use of fertilizers, increasing area under high yielding varieties and enhancing available irrigation potential.

Landes, (2004) has analysed that India's high agricultural tariffs and growing farm subsidies has received much attention, on the other hand implications of domestic policies that has discouraged agricultural investment and created inefficient domestic markets has received less attention. In this study, the author used an approach for analysing the effects of improvements in marketing efficiency that might be associated with reforms that strengthening the climate and incentives for private investments in agriculture. The author compared these results with the impacts of more traditional reform scenarios involving the elimination of agricultural subsidies and tariffs. Due to the equity implications of reforms are a key consideration for Indian policy makers, the author used a framework that disaggregated the household sector between rural and urban and by income class. The author suggested that in contrast to subsidy and tariff reforms measures leading to improved agricultural marketing efficiency could yield substantial economy wide gains in income and employment as well as positive price impacts for both producers and consumers and distributional gains favouring low income households.

Acharya, (2004) explained that the genesis of input subsidies in Indian agriculture can be traced to the philosophy and objectives of agricultural development strategy launched during the mid-1960s. Input subsidies help in balancing the conflicting interests of farmers and consumers and in achieving macro food security. Subsidies on fertilizers, electricity and canal water, which account for bulk of subsidies has been analysed. In 1999-2000, the electricity subsidy accounted for 53 per cent, fertilizers subsidy was 28 per cent and 19 per cent was canal irrigation subsidy. During the last twenty years, 81 per cent of the incremental subsidy has been contributed by increase in the rate of per unit subsidy, contrary to general perception, Punjab has accounted for only 7.4 per cent of the total subsidies in Indian agriculture Across farm size groups distribution of subsidies has been to follow the pattern of percentage share of operated areas. Crop wise analysis has revealed that the input subsidies are mainly going to the food crops. The author suggested that should be handled cautiously the issue of subsidies in Indian agriculture because the economic conditions of farmers have not improved to a desirable level. Subsidies on farm inputs cannot be seen in isolation of the subsidies in other sectors of the economy, which were many times more, and consequences of their withdrawal were less painful.

Rao, (2005) conducted a study to estimate the extent of subsidy to fertilizes, power and canal irrigation in Andhra Pradesh. The author used data from the year 1980-81 to 2002-03. The author found that total subsidies in Andhra Pradesh have increased from Rs. 167.8 crores in 1980-81 to Rs. 4,100.4 Crores in 2002-03. The author used Cobb-Douglas production function method to find out the percentage share of agricultural subsidies during 2002-03 in three different regions of Andhra Pradesh *i.e.* coastal Andhra, Rayalaseama and Telengana regions. The author found that coastal Andhra received high percentage share of fertilizer and irrigation subsidy than other two regions and Telengana received high percentage share of power subsidy during 2002-03. The author also found that during 2002-03, total agricultural subsidies per hectare of irrigated land was Rs.13,826 and Rs.397 per hectare of rain fed land. The author suggested that simultaneous reform of all the three subsidies is needed and equity between irrigated and rain fed farmers to be addressed and many fold investments needed when subsidies be reduced.

Morris, (2005) stated that subsidy on irrigation has caused considerable distortion on the cropping pattern in Punjab. Majority of the cultivable land has become mono-crop culture and has favoured water– intensive crops like paddy, subsidy on canal water and electricity has led to excessive irrigation causing salinity adapter logging in some areas. Subsidies on fertilizers have led to excessive application with adverse environmental effects. The subsidies on the production of fertilizer have not been even on all fertilizers. The main prominent feature was the excessive consumption of Nitrogen based fertilizers in both the seasons. This is so because of the lower cost of Nitrogen based fertilizers as they were more subsidized than the other fertilizers based on potash and phosphorus. There are also wide regional variations in the consumption of fertilizers. In 1999-2000, the preparation of NPK consumption in Mansa district of Punjab was 378:122:1 as compared to that of Jalandhar district which recorded 9:35:1. Intensive cultivation of paddy and wheat is causing nitrate pollution and excessive chemicalization of soils and rising deficiency of micronutrients. Application of chemicals to the soils can cause irreversible damage to the content of the soil, killing its microorganisms that are essential or maintaining the soil structure and basic fertility. The water pricing policy was only aggravating the problem. The policy of providing free electricity for agricultural purpose was causing indiscriminate use of electric motors and excessive water use. This is one of the reasons for depletion of the ground water table. The author suggested that government stared the farmers for balance use of subsidies.

Singh, (2005) examined the issue of equity in fertilizer subsidy distribution in India, in terms of percentage share of different farm classes, crops and states in total fertilizer use as well as per hectare fertilizer use on different size categories of farms. Date has been mainly taken from the Agricultural Input Survey carried out along with the Agricultural Census of 1991-92. The study showed that paddy and wheat cultivators were the major beneficiaries of fertilizer subsidy. Inter-state disparity in fertilizer consumption still remains high, though it has been falling over the years. The study suggested that a uniform approach to reduction of all types of subsidies is not justified. Instead, a well thought out, properly regionally differentiated approach to subsidy reduction needs to be adopted.

Modi, (2006) observed that there is no denying the fact that the expenditure incurred on irrigation soil conservation and agricultural research will be more productive than expenditure of the same amount being doled out as subsidies. The author found that public investment is agriculture and allied sectors in 1985-86 at constant prices were of the order of Rs. 6,213 crores. In 1993-94, it amounted to Rs. 4,918 crores on the other hand, total agriculture subsidies were Rs. 14069 crores. There was some increase in the public investment during 1994-95. Since then, it has again fallen from Rs. 5,369 crores to Rs. 4,658 crores in 2001-02 amounted to Rs. 36,224 crores. These figures clearly indicated that a part of the enormous increase in the subsidies paid to the agricultural sector were at the cost of public investment in agriculture. The author argued that if the government did not withdraw the subsidies, it will obviously try to reduce the fiscal deficit by cutting down its expenditure on some other count, a cut which may be more harmful for the economy than the withdrawal of subsidies.

Chandrashekhar, (2006) explained that subsidies given by Organization for Economic Co-operation and Development (OECD) countries, especially the United State and the European Union are huge. Every ton of commodity produced is subsidized to the extent of between 25 per cent and 50 per cent of the market price. These subsidies encourage more production, augment global supplies, depress would prices and di-start the discovery of free market prices. Therefore, a strong move to pressure the developed economies to phase out farm support. The foregoing simple analysis shows that in none of the four major commodities would India stand to benefit substantially if the subsidies were eliminated. It may be politically correct and perhaps expedient for India to make appropriate noises against farm subsidies at global forms such as the World Trade Organization. While reduction or elimination of subsidies would impact world commodity prices, consuming and importing nations would be the worst hit. Unlike several countries, there are dependent on farm goods export, India is a large consuming country, so subsidy induced low prices would be in Indian Consumer's interest. The author explained the agriculture situation in India that, a lot of attention is now being showered on the farm sector but given the neglect of this sector for decades, entrenched problems and serious changes in raising production and productivity, self-sufficiency itself would be a distant dream, leave alone generating export surplus of major commodities. Developed economies can afford huge payment as subsidy. India cannot concert efforts are required to strengthen agriculture, improve agro-produce marketing and enhance quality. The author suggested that huge investments are necessary for this national effort. Government will have to find adequate financial resources for strengthening input delivery system, expanding irrigation, building rural infrastructure and delivering price/market information to farmers.

John, (2006) argued that on the one hand there is a need to get rid of the subsidies, on the other if they were all withdrawn at once thousands of farmers would go out of business and food prices would really go through the roof. Farm subsidies came in with Roosevelt's New Delhi Policies the same thing that gave us welfare payments to the poor. Today, both systems have grown completely out of control. Agricultural subsidies are geared to help farmers keep their production cost low and Governments cannot cut all welfare because hundreds of thousands of people would starve. Both systems are going to have to be drawn down gradually until hopefully they can be done away with altogether without harm to people and our country. The author suggested that there should be a shift in the way subsidies are handed out, that would help small size category farmers using sustainable farming practices to make a decent living off of the land. This would tend to encourage more diversity of crops, "better" (in the sense of more sustainable) use of farm land and higher quality of production available to the consumer.

Pachauri, (2006) pointed out that past election, Punjab has announced to implement the provision of free electricity for farmers and for some other sections like scheduled caste and below poverty line consumers. This policy of free electricity is imposing additional financial burden on the Punjab Government. However, free power to farmers, leads to installation of in-efficient pump sets, which use excessive energy, wastage of energy, for given output. Therefore, if India has to attain a level of economic success globally, then a strong policy to install power stations is an essential pre-requisite and urged the Prime Minister of India putting an end to politicians promising free electricity to the farmers which has not remained a demand of farmers.

Jain, (2006) made an attempt to analyse the provision of agricultural subsidies, which have burdened Punjab's exchequer heavily. This study highlighted the existence of disparities in the flow of electricity subsidy between the progressive and backward areas. The author conducted a primary survey in two districts *viz.* Mansa and Ludhiana to make a comparative study of the flow of electricity subsidy to different classes of the farmers. The sample size was of 300 farm household and multistage sampling technique was adopted. The author found an association between the area and the availability of electricity connections through a chi-square test. The results showed that the proportion of farmers having electricity connections in the progressive area was 51 per cent higher than the backward areas. The positions in both areas were almost similar regarding electricity connection ownership by medium and large size category farmers, but there exist differences between the two regions in case of small size category farmers. In the backward area, a very small proportion (3.7 per cent) of the farmers own electricity connections. In the progressive area, it has been very high (79.4 per cent). The author observed the farmers in progressive area were getting electricity for a minimum of eight hours and uninterrupted electricity supply, bid farmers in backward area were getting electricity between six top eight hours and interrupted electricity supply. The author also observed that the provision of electricity subsidy has a negative impact on the sustainability of agriculture as it has implications for depletion of underground water. The availability of this cheap mode of irrigation also persuaded farmers in the backward area to shift to cultivation of water intensive crops like rice, from their old practices of growing cotton. The author also observed that the state's policy of providing electricity subsidy to the agricultural sector has weekend the financial soundness of the Punjab State Electricity Board. This study pointed out that a majority of the farmers, who were unsatisfied with inadequate, unreliable electricity supply under the subsidy regime were willing to pay reasonable user charges. Though some farmers were favouring flat rates, a majority agreed to accept metered rate in both areas, provided the availability of quality electricity is ensured. The study also observed that economic incentive for the farmers to make a choice between diesel-based irrigation and electricity-based irrigation. On the basis of this evidence, the author put forward the case for user charges-based open access to electricity to speed up the pace of economic development of an agro-based economy as this policy, apart from bringing hope for the sustainability of the electricity utility, will ensure enough economic returns to the farmers depended on non-electrical means of irrigation.

Anderson, (2007) estimated the impacts of all merchandise trade distortions (including agricultural subsidies) globally by using the linkage model of the global economy (Projected to 2015). Results suggested that developing countries' economies bear a disproportionate burden of current distortions, reducing their average income by 0.8 per cent compared with 0.6 per cent for high-income countries. A huge 63 per cent of those costs are due to agricultural market distortions, even though agriculture accounts for just 4 per cent of global Gross Domestic Product. As much as 93 per cent of the cost of these agricultural distortions is due to import barriers and only 2 per cent to agricultural export subsidies and 5 per cent to direct domestic support programmes. Half of the overall cost of developing countries is due to the region's own policies, partly because they trade with each other fairly intensively and partly because their own trade barriers are higher those of high-income countries. The author concluded that if all those trade distorting measures were to be removed, the developing countries' percentage share of global output as of 2015 would rise from 70 to 75 per cent for primary agricultural goods and of textiles and clothing from 62 to 65 per cent. Developing country's percentage share of global exports would rise even more dramatically, especially in agriculture from 47 to 62 per cent in primary farm products and from 34 to 40 per cent in processed farm products that represents a rise in developing country exports of around $ 200 billion per year (in 2001 US dollars) - an increase of two-thirds compared with baseline scenario for 2015 and in exports of non-agricultural goods of $ 400 billion per year. The author suggested that multilateral cuts in tariff bindings are especially helpful because they can lock in previous unilateral trade liberalizations and they can be used as an opportunity to multilaterals previously agreed preferential trade agreements and thereby reduce the risk of trade diversion from those bilateral of regional agreements.

Gulati, (2007) reviewed the trends in government subsidies and investments in and for Indian agriculture. He developed a conceptual framework and model to assess the impact of various subsidies and investments on agricultural growth and poverty reduction and also presented several reform options with regard to re-prioritizing government spending and improving institutions and governance. There are three major findings: (i) Initial subsidies in fertilizer and irrigation have been crucial for small size category farmers to adopt new technologies. Small size category farmers are often losers in the initial adoption stage of a new technology since prices of the agricultural products are typically being pushed down by greater supply of products from large farms, which adopted the new technology. But as more and more farmers have adopted High Yield Varieties, continued subsidies have led to inefficiency of the overall economy. (ii) Agricultural research, education and rural roads are the three most effective public spending items in promoting agricultural growth and poverty reduction during all periods. (iii) The trade-off between agricultural growth and poverty reduction is generally small among different types of investments. As for agricultural research, education and infrastructure development, they have growth impact and a large poverty reduction impact. In this study several policy lessons are drawn by the author. Agricultural input and output subsidies have proved to be unproductive, financially unsustainable, and environmentally unfriendly in recent years and contributed to increased inequality among rural Indian States. The author suggested that to sustain long-term growth in agricultural production and therefore provide a long-term solution to poverty reduction, the government should cut subsidies of fertilizer, irrigation, Power and credit and increase investments in agricultural research and development, rural, infrastructure and education. Promoting non-farm opportunities are also important.

Jha, (2007) explained that one of the principle elements of the economic reforms program initiated in 1991 was to reduce the fiscal deficit of the central Government which, at that time, faced a solvency crisis. This reduction was at least partially achieved by reducing transfers to state governments. As a result, state government budgets faced crisis and agriculture being largely a state subject, was denied adequate investment. This study reviewed the performance of Indian agriculture, particularly in the post – reform period. Thus, while current operations are being subsidized to some extent resource for augmentation of productive capacity in agriculture are dwindling. The author further analysed that the current stance of policy towards Indian agriculture is neither efficient nor equitable. The stagnation of agricultural investment has meant that enough productive capacity to sustain agricultural growth has not been forthcoming. At the same time political economy considerations have led to a burgeoning of the agricultural subsidies bill. The subsidy mix has not been well thought out and more importantly, the subsidies are available for current production and not addition for productive capacity. Furthermore, there is widespread evidence that the more efficient farmers are able to gain a disproportionately large part of the subsidies, the subsidy incidence is inequitable. At the same time, the stagnation of agriculture has led to a spill over of problems into other areas particularly, but not exclusively, in the area of unemployment. Indian agricultural subsidies, though at historic high levels, are low when compared to European on US levels. The author suggested that there is an urgent need to put into effect on expenditure switch from subsidies to investment to lift Indian agriculture from its current stagnation.

Birner, *et al.* (2007) analysed the political economy of agricultural subsidy reform in India, taking the case of electricity subsidies for groundwater irrigation as an example. The study is motivated by the recent debate on the question of whether agricultural subsidies can be a useful strategy to promote food production and agricultural development. India is an interesting case, subsidies were used effectively to promote the Green Revolution, but reforming agricultural subsidy policies proved to be politically difficult once the subsidies has fulfilled their original purpose and started to cause fiscal, distributional and environmental problems. The study is based on an explorative analysis of a data set that captures the variation between electricity subsidies for groundwater irrigation across states and over time. An illustrative case study of the state of Andhra Pradesh complements this analysis. The author analysed that an interest group approach that focuses only on the role of better-off farmers in preventing subsidy reform is too limited to capture the complex nature of the politics of subsidy reform. The author suggested that a wider range of factors, including path dependency, party ideology and electoral politics, need to be considered to understand the challenges of subsidy reform and identify promising reform strategies.

Sharma, (2007) stated that with paddy and wheat, heavily water dependent crops, farmers have every reason to over-exploit ground water. The inevitability of ground water extraction has been politically exploited too. Successive governments in the recent past have even given free electricity to the farmers in the state. The water tables have fallen at alarming rates in many places in the state during the last few decades. The government's policy of providing free electricity for agriculture and very low water charges for canal water has encouraged inefficient use of irrigation water. Intensive use of tube well irrigation has led to depletion of water resources in the state. The author found that about 98 per cent of ground water resources in the state have already been exploited. Nearly 59 per cent of blocks in the state have overexploited ground water resources, the highest rate in the country and another 12 per cent rate in dark/critical zone. On the other hand, injudicious use of canal irrigation water without regard to soil conditions and inadequate attention to drainage, has led to the emergence of conditions of water logging and salinity in many areas resulting in valuable agriculture land going out of use in the state. The author suggested that free electricity should be withdrawn for efficient use of water.

Jakhar, (2008) tried to analyse that who so ever has been in politics would know that subsidies don't benefit anyone in the long run. A populist promise is a one shot affair, which can be enchased just once. It only raises the expectations of the common man who wants more in the next elections. Instead of realizing this truth, it is a shame that parties in the state are engaged in populist one-up-man ship. The author observed that free power to farmers is pushing big financial burden on the Punjab State Electricity Board. The author observed that the farmer is not afraid of paying up, provided he gets the service he is being charged for. The author suggested that the subsidies should be replaced with constructive schemes that empower people and give them that one push they need to get out of poverty.

Dhillon, (2008) argued that it is unfortunate that politicians and bureaucrats without a vision are driving the state (Punjab) to its ruin all because they cannot plan a new paradigm for development. Politicians seem to think subsidies are the only way out to gain votes, thus limiting their horizon to five years. Subsidies in the form of seeds, fertilizers etc. are provided even in the most developed countries and Punjab cannot be an exception. But power subsidy needs to be a streamlined. Free power should be restricted to small and medium landholders. This should be further distributed according to the crop and there should be no free electricity for paddy. The author suggested that the government needs to give a push to education, which is going from bad to worse. The State Government can always mop up money for this initiative by extending the house tax to rural areas. If people in villages can pay for petrol or diesel, why not for house tax. The rural areas must be provided with basic amenities like drinking water, sewerage, good roads and schools to stop migration to cities. The author analysed that when the government itself is wasting money, how it can teach people to save it. Instead of these schemes, the government should focus on just three things - electricity generation, infrastructural development and water supply. The accompanying development will take care of the rest.

Virk, (2008) described the issue of un-remunerative minimum support price (MSP) and the agricultural subsidies, in Punjab State. The issue of un-remunerative MSP has been as contentious one for the past many decades and a majority of the farmers agitations revolved around it from time to time. Given the fact that 59 per cent of the farmers in our country plough less than 2.5 acres of land and another 32 per cent between 2.5 and 10 acres, it becomes evident that being small and marginal farmers they do not enjoy the economies if the scale and hence have more input cost. Thus, the pricing should be done keeping in view their cost of production and not by taking other theoretical factors. Another anti-farmers stance if the government is the issue of subsidies, which are not directly given to the farmers but circumvented a very little of which reaches him and benefits the others more. The proposal of giving fertilizer subsidy direct to farmer, which has gone to Rs. 90,000 crores this year, six times up from Rs. 15,799/- crores in 2004-05, by the union FM was shot down by none other than the chemical and Fertilizers Ministry, obviously, to benefit the fertilizer lobby at the cost of the farmers. In the year 2007-08, the Government procured, 22.44 million tons of wheat and 44.6 million tons of paddy, for which the total payment made to the farmers of the country based on the declared MSP works out to be Rs. 54,106 crores. If the huge amount of fertilizer subsidy is diverted to farmers, they hugely benefit even if the rates of fertilizers are hiked two to three times. The author suggested that considering these facts, it is time that government should seriously think to formulate a farmer's friendly agro-policy by rationalizing subsidies and making the CACP (Commission for Agricultural Costs and Prices) an independent, broad –based and transparent body instead of keeping it infested with textbook economists and bureaucrats. Such a policy should ensure remunerative prices to farmers, which in turn, will add to the growth rate of the country and ensure the much needed national food security.

Singh, (2008) stated that the truth is that though everything is done in the name of the farmers, the benefits do not reach them. Since 1997, the farmers of Punjab are being provided free electricity and water on paper, but the reality is that the farmers hardly use electricity for 65 days in a year, and that too very erratically. The Punjab State Electricity Board (PSEB) that is always griping about being overburdened admits that 24 pc of electricity is lost during transmission while 10 pc is stolen, both of which can be plugged. In the period between 1997 and 2007, Punjab has incurred losses to the tune of Rs. 30,000 crores due to Punjab State Electricity Board. This led to paucity of funds for other departments and development work suffered heavily. It is a known fact that directly appointed officers has a better vision than the promoters. The development of Punjab cannot be done by a few honest civil servants. It is of utmost importance that the Agriculture Minister should be highly educated and well versed with farming practices. The author suggested that the Punjab Government should stop giving all these subsidies and instead provide financial aid for education of the children of the farmers and farm workers directly into their bank account. They should also be given free health insurance cover.

Chand, *et al.* (2008) showed the distribution of fertilizers subsidies in different states of India. In this study, the author showed that out of total subsidy on fertilizer in the country, largest percentage share (18.1 per cent) goes to Uttar Pradesh followed by Andhra Pradesh (11.41 per cent). Around 9 per cent of total subsidies go to Maharashtra and Punjab each. Percentage share of Assam, Himachal Pradesh, Jammu & Kashmir and Uttarakhand was below 1 per cent. But this distribution does not indicate that which state benefits more from subsidies because of variation in the size of State. The author calculated the per hectare subsidy and found that fertilizer subsidy on per hectare basis varies in the range of Rs.393 in Rajasthan to Rs. 3,167 in Punjab. After Punjab, the second most benefited State is Haryana with subsidy of Rs. 2,516 per hectare of net sown area. Farmers in West Bengal, Uttar Pradesh and Andhra Pradesh are estimated to get per hectare subsidy between Rs. 1,626 and Rs. 1,730. Among other States, per hectare subsidy was above Rs. 1,000 in Uttarakhand, Bihar and Tamil Nadu. States which less than Rs. 600 subsidy are Assam, Chhattisgarh, Jharkhand, Madhya Pradesh, Odisha and Rajasthan. The author suggested that more fertilizer subsidy should be given to poor states to remove the imbalance use of fertilizer.

Pandey, (2008) tried to examine the impact of fertilizer subsidy on food grain production in India. For this purpose, the author showed relationship between fertilizer price and food grains productions and two equations were estimated simultaneously using SURE technique of regression analysis by using statistical package EVIEWS. The estimates were based on data for the period 1980-81 to 2004-05. The author showed that if subsidy on fertilizer is removed completely then price of fertilizer increased by 69 per cent and this would cause close to 9 per cent reduction in food grain production in the country. Then author concluded that if subsidy on fertilizer is taken away in one go it is going to cause very serious adverse effect on food grain production and consequently on food security. The author suggested that there is a need to keep a check on growth of fertilizer subsidy without causing adverse effect on food grain production is to increase prices of fertilizer by suitable fraction of increase in food grain prices.

Maurice, *et al.* (2008) analysed that India's high agricultural tariffs and growing farm subsidies have received much attention, whereas the implications of domestic policies that have discouraged agribusiness investment and created inefficient domestic markets have received less attention. The authors devise an approach for analysing the effects of improvements in marketing efficiency that might be associated with reforms that strengthen the climate and incentives for private investment in agriculture and agribusiness. These results are compared with the impacts of more traditional reform scenarios involving the elimination of agriculture subsidies and tariffs. The equity implications of reforms are a key consideration for Indian policymakers, a framework used that disaggregates the household sector between rural and urban and by income class. The results suggest that in contrast to subsidy and tariff reforms, measures leading to improved agricultural marketing efficiency could yield substantial economy-wide gains in income and employment as well as positive price impacts for both producers and consumers and distributional gains favouring low income households.

World Development Report, (2008) stated that faster agricultural growth and increased response to better price incentives depend on investments in crores, public goods such as market infrastructure, research institutions and support services. In most of the countries, public investments were low and public to budget allocations to subsidies were high *i.e.* 37 per cent in Argentina, 43 per cent in Indonesia, 75 per cent in India and 75 per cent in Ukraine, 26 per cent in Kenya during 2002-03. This report showed that big portion of these subsidies went to larger farmers. In India, subsidies have crowded out investments in crores for public goods or investment, which has fallen. This study concluded that more and better quality expenditures require improved budgetary process aligned to well-articulated agricultural strategies. Greater public disclosure and transparency of budget allocation and impacts are needed to mobilize political support for budgetary reforms.

Halmandage, (2009) stated that during the last decade subsidies provided by government of India have grown at a very rapid rate. The subsidies rose from 1.7 per cent of total budget expenditure in 1970-71 to more than 10 per cent in 1980-81. Agricultural subsidies and food subsidies constituted about 17 per cent of the total subsidies in country. Substantial additional growth in agricultural production is needed to meet the basic necessities of large and growing population. It is also needed to generate agricultural surplus required for economic development with emphasis and employment equity. The bulk of growth agricultural production will have to come from continuous increase in the productivity of land. Yield based growth cannot sustain without removing soil fertility constraints and promote technological changes. For both these process substantial growth in fertilizer use is necessary. After the cut of the fertilizer subsidy, production cost of agricultural will increase. The decrease in agricultural production will affect the various sectors.

This will increase the problems on unemployment, poverty, inequitable distribution of income and unbalanced growth and trade in various regions. The government subsidy policy protects the weaker sections of consumers and marginal farmers. The change in the policy of fertilizer subsidy, will affect the weaker sections of consumers and marginal farmers.

Bhalla, (2009) tried to analyse the performance of agriculture at the state level in India during the post-reform period (1990-93 to 2003-06) and the immediate pre-reform period (1980-83 to 1990-93) shows that the post-reform period has been characterised by deceleration in the growth rate of crop yields as well as total agricultural output in most states. By ending discrimination against tradable agriculture, economic reforms were expected to improve the terms of trade in favour of agriculture and promote its growth. The study also discusses the cropping pattern changes that have taken plashes in area allocation as well as in terms of value of output. The slowdown in the process of cropping pattern change means that, most of the government's effort to diversify agriculture, have failed to take off. It is beyond the scope of this study to undertake a comprehensive analysis of the main reasons for the failure of economic liberalisation to improve the state of agriculture in India. But, it is hoped that the state and region wise analysis of agricultural growth during the pre- and post-liberalisation period undertaken above would provide a backdrop to scholars and policymakers to undertake an in-depth analysis of the reasons for slowdown in agriculture in the post-reform period.

Sud, (2009) proposed the policy of direct transfer of fertiliser subsidy to farmers is misconceived and inappropriate. The author analysed that per-hectares subsidy on marginal farms is more than double that on large farms. The average subsidy was the highest (Rs. 916.2 per hectare) on marginal farms and the lowest (Rs. 405.8 per hectare) on large size category farmers. The percentage share of marginal farmers in the total fertiliser subsidy was the highest (28.3 per cent), followed by small farms (23 per cent). It was the lowest (6.3 per cent) on large farms. However, state-wise distribution of fertiliser is skewed. More than half of the total fertiliser subsidy is cornered by five top fertiliser consuming states-Uttar Pradesh, Andhra Pradesh, Maharashtra, Madhya Pradesh and Punjab. These states grow fertiliser-intensive crops, such as rice, wheat, cotton and sugarcane. But, over a period of time, the percentage share of these states in the total fertiliser use is declining. It fell from about 60 per cent in 1992-93 to 55.8 per cent in 1999-2000 and to 54.5 per cent in 2007-08. The other major beneficiary states are Gujarat, Karnataka, West Bengal, Bihar, Haryana and Tamil Nadu. Their percentage share in total fertiliser subsidy increased from 31.7 per cent in 1992-93 to 36 per cent in 2007-08. Among the crops, rice and wheat are major consumers of fertilisers and account for over half the total subsidy. Rice is the biggest beneficiary of fertiliser subsidy, receiving 32.2 per cent of the total in 2001-02. Wheat was next with a 20.3 percentage share of fertiliser subsidy, followed by sugarcane (6.3 per cent), cotton (5.9 per cent). Justifying the continuing of the fertiliser subsidy and the mode of its transfer to farmers, the author has concluded that a reduction in the subsidy is likely to have an adverse effect on farm production and income of small and marginal farmers, as they do not benefit from higher output prices but do benefit from lower input costs.

Sharma, *et al.* (2009) tried to examine the trends in fertilizer subsidy and the issue of distribution of fertilizer subsidies between farmers and fertilizer industry, across regions/states, crops and different farm sizes. There is a general view in academic, policy and political circles that agricultural subsidies are concentrated geographically, they are concentrated on relatively few crops and few producers and in many cases do not reach the targeted group(s). The author analysed that fertilizer subsidy has increased significantly in the post-reforms period from Rs. 4,389 crores in 1990-91 to Rs. 75,849 crores in 2008-09. As percentage of GDP, this represents an increase from 0.85 per cent in 1990-91 to 1.52 per cent in 2008-09. The fertilizer subsidy is more concentrated in few states, namely, Uttar Pradesh, Andhra Pradesh, Maharashtra, Madhya Pradesh, and Punjab. Inter-state disparity in fertilizer subsidy distribution is still high though it has declined over the years. Rice is the most heavily subsidized crop followed by wheat, sugarcane and cotton. These four crops account for about two-third of total fertilizer subsidy. The study highlighted the existence of fair degree of equity in distribution of fertilizer subsidy among farm sizes. The small and marginal farmers have a larger percentage share in fertilizer subsidy in comparison to their percentage share in cultivated area. A reduction in fertilizer subsidy is, therefore, likely to have adverse impact on farm production and income of small and marginal farmers as they do not benefit from higher output prices but do benefit from lower input prices.

Badiani, *et al.* (2010) stated that states in India received the authority to set electricity prices, electricity pricing emerged as a powerful political tool and politicians promised subsidized electricity prices for agriculture in return for the agricultural sector's vote. While these subsidies reduced rural poverty and increased agricultural productivity, they may have also encouraged the overexploitation of groundwater resources and encouraged the exit and downsizing of the industrial and commercial sectors. In this study it is evaluated whether subsidized electricity prices increased groundwater extraction and over-exploitation, using panel data from 265 districts (the U.S. equivalent of a county) in India from 1997 to 2004. To isolate the effect of electricity prices on groundwater extraction, the author have taken two strategies - an instrumental variables model where the authors used pre-scheduled state electoral cycles as an instrument for electricity prices and a model where the authors estimated the differential effect of electricity subsidies on districts characterized by different groundwater prices. This study found that a 25 per cent increase in the agricultural price of electricity will reduce groundwater extraction by roughly 0.7 to 2.1 per cent. However, if agricultural electricity prices were set equal to industrial electricity rates demand would drop by 23 to 67 per cent.

Halmandage, *et al.* (2010) stated that subsidies are among the most powerful instrument for manipulating or balancing the growth rate of production and trade in various sectors for an equitable distribution of income for protection of the weaker sections of the society. The support and procurement prices of major agricultural production are some of the important measure which is done to protect the interest of farmer and weaker section of consumers. Substantial additional growth in agricultural production is needed to meet basic necessities for a larger growing population. It is also needed to generate agricultural surplus required for economic development with emphasis on employment equity. The agricultural production increased in initial period gradually after that the fertilizer subsidies were reduced, the overall economy affected. The government policy of subsidy is mainly for protection of weaker sections and marginal farmers. If the Government of India reduces the subsidy of fertilizer, it will affect the overall economy, agricultural production, equity of give me to regional imbalance, problems like employment and poverty.

Grossman, *et al.* (2011) stated that in India, agricultural trade policy is a part of a larger food and agriculture policy regime that seeks to maintain food self-sufficiency while providing income support to the agricultural sector and poor consumers. The Government of India (GOI) uses a variety of policy instruments to achieve these goals, including domestic subsidies to inputs, outputs, transportation, storage and consumption to reduce producer costs and consumer prices and broader measures such as subsidies, tariffs, quotas, and non-tariff measures to protect domestic producers from import competition, manage domestic price levels, and guarantee domestic supply. Input subsidies are the most expensive aspect of India's food and agriculture policy regime, requiring steadily larger budget percentage share. India subsidizes agricultural inputs in an attempt to keep farm costs low and production high. GOI's intended result is for farmers to benefit from lower costs, but also for them to pass some of the savings on to the consumers in the form of lower food prices. These policies result in effective subsidies to the farmer of 40 to 75 per cent for fertilizer and 70 to 90 per cent for irrigation and electricity. Input subsidies can also produce unintended effects. According to GOI reports, input subsidies have resulted in overutilization of inputs. This overutilization has in turn led to soil degradation, soil nutrient imbalance environmental harm, and groundwater depletion, all of which have caused declined effectiveness of inputs. Additionally, input subsidies distort trade by increasing net exports of input intensive commodities while decreasing net exports of commodities which require relatively fewer inputs.

From the above studies, it may conclude that agriculture subsidies are a worldwide phenomenon. Some studies showed the distribution pattern of agriculture subsidies in different countries and in different states of India. Whereas some studies showed the impact of agriculture subsidies on income of farmers of different states of India, on agriculture production, on gross cropped area, on cropping pattern etc.

## Chapter 3

# Gross Cropped Area in India

Subsidies are among the most powerful instruments for manipulation or balancing the growth rate of production and trade in various sectors and regions and for an equitable distribution of income for the protection of weaker sections of society. The support and procurement prices for more agricultural production are some of the important measures, which are done to protect the interests of farmers. During the last decade subsidies provided by government of India have grown at a very rapid rate. The subsidies rose from 1.7 per cent of total budget expenditure in 1970-71 to more than10 per cent in 1980- 81. Agricultural subsidies and food subsidies constituted above 10 per cent of the total subsidies in country (Halmandage, 2010).

Land is the fundamental basis for the most of the human or natural activities and is one of the major natural resources on earth. Agricultural productivity is entirely dependent on the availability of suitable land (State of Environment Punjab, 2007). In India, there are competing demands of area available for cultivation from increase in rural habitations, forestation, urbanisation and industrialisation. Consequently, gross cropped area in the country has registered a rapid deceleration in its growth over time (Bhalla, 2009).

In this chapter, an attempt is made to analyse the gross copped area (GCA) during 1980-81 to 2006-07.

The gross cropped area (GCA) in India during 1980-81 to 2006-07 is shown in Table 3.1. This table reveals the west zone got topmost position, followed by north zone, south zone, east zone and north-east zone throughout the study period. In India, GCA has shown variations *i.e.* it has increased from 1,73,324 thousand hectares in 1980-81 to 1,85,403 thousand hectares in 1990-91 and further increased to 1,88,601 thousand hectares in 1996-97, it has declined to 1,86,565 thousand hectares in 2000-01 and further declined to 1,75,678 thousand hectares in 2006-07. As zone-wise analysis shows that in west zone, the GCA has increased from 69,882 thousand hectares in 1980-81 to 75,659 thousand hectares in 1990-91 and further increased to 78,097 thousand hectares in 1996-97 and declined to 72,833 thousand hectares in 2006-07.

**Table 3.1: Zone-wise Distribution of Gross Cropped Area in India during 1980-81 to 2006-07 (In 000 hectares)**

| *Years/Zones* | *1980-81* | *1985-86* | *1990-91* | *1996-97* | *2000-01* | *2006-07* |
|---|---|---|---|---|---|---|
| South | 32,363<br>(18.67) | 33,054<br>(18.62) | 34,688<br>(18.71) | 35,333<br>(18.73) | 35,271<br>(18.91) | 36,368<br>(20.70) |
| West | 69,882<br>(40.32) | 71,628<br>(40.35) | 75,659<br>(40.81) | 78,097<br>(41.41) | 75,231<br>(40.32) | 72,833<br>(41.46) |
| North | 38,806<br>(22.39) | 39,918<br>(22.49) | 40,969<br>(22.10) | 42,132<br>(22.34) | 43,233<br>(23.17) | 39,780<br>(22.64) |
| East | 27,514<br>(15.87) | 27,763<br>(15.64) | 28,741<br>(15.50) | 27,416<br>(14.54) | 27,043<br>(14.50) | 20,246<br>(11.52) |
| North-East | 4,759<br>(2.75) | 5,163<br>(2.91) | 5,346<br>(2.88) | 5,623<br>(2.98) | 5,787<br>(3.10) | 6,451<br>(3.67) |
| India | 1,73,324<br>(100) | 1,77,526<br>(100) | 1,85,403<br>(100) | 1,88,601<br>(100) | 1,86,565<br>(100) | 1,75,678<br>(100) |

*Source*: Government of Punjab, Statistical Abstract, Various Years.

*Note*: Percentages are shown in parentheses.

In north zone, it has increased from 38,806 thousand hectares in 1980-81 to 42,132 thousand hectares in 1996-97 and declined to 39,780 thousand hectares in 2006-07, whereas in south zone, the GCA has increased from 32,363 thousand hectares in 1980-81 to 34,688 thousand hectares in 1990-91 and further increased to 35,333 thousand hectares in 1996-97 and declined to 35,271 thousand hectares in 2000-01 and again increased to 36,368 thousand hectares in 2006-07.

In east zone, the GCA has increased from 27,514 thousand hectares in 1980-81 to 28,741 thousand hectares in 1990-91 and declined to 27,416 thousand hectares in 1996-97 and further declined to 20,246 thousand hectares in 2006-07, on the hand the GCA has increased from 4,759 thousand hectares in 1980-81 to 5,163 thousand hectares in 1985-86 and further increased to 6,451 thousand hectares in 2006-07 in north-east zone.

It reveals that at national level, the GCA in absolute terms has increased during pre-liberalisation period, whereas declined during post-liberalisation period. West zone has got topmost position by receiving maximum percentage share followed by north, south, east and north-east zones during the study period.

As percentage-wise analysis shows that the percentage share of west zone has increased from 40.32 in 1980-81 to 41.41 in 1996-97 and declined to 40.32 in 2000-01, the percentage share of north zone has declined from 22.39 in 1980-81 to 22.10 in 1990-91 and increased to 23.17 in 2000-01. The percentage share of south zone has increased from 18.67 in 1980-81 to 20.70 in 2006-07 on the other hand in east it has declined from 15.87 in 1980-81 to 14.54 in 1996-97 and further declined to 11.52 in 2006-07. The percentage share has increased from 2.75 in 1980-81 to 2.88 in 1990-91 and further increased to 3.67 in 2006-07 in north-east.

Tamil Nadu and Kerala. It is observed that in Andhra Pradesh, the GCA has declined from 12,281 thousand hectares in 1980-81 to 12,100 thousand hectares in 1985-86 and increased to 13,410 thousand hectares in 1996-97, it has again declined to 12,811 thousand hectares in 2006-07.

**Table 3.2: State-wise Distribution of Gross Cropped Area in South Zone in India during 1980-81 to 2006-07 (In 000 hectares)**

| *Years/States* | *South Zone* | | | | | |
|---|---|---|---|---|---|---|
| | *1980-81* | *1985-86* | *1990-91* | *1996-97* | *2000-01* | *2006-07* |
| Andhra Pradesh | 12,281<br>(7.09) | 12,100<br>(6.82) | 13,192<br>(7.12) | 13,410<br>(7.11) | 13,545<br>(7.26) | 12,811<br>(7.29) |
| Karnataka | 10,660<br>(6.15) | 11,146<br>(6.28) | 11,759<br>(6.34) | 12,335<br>(6.54) | 12,284<br>(6.58) | 12,438<br>(7.08) |
| Kerala | 2,862<br>(1.65) | 2,866<br>(1.61) | 3,020<br>(1.63) | 3,020<br>(1.60) | 3,022<br>(1.62) | 2,918<br>(1.66) |
| Tamil Nadu | 6,469<br>(3.73) | 6,819<br>(3.84) | 6,632<br>(3.58) | 6,457<br>(3.42) | 6,338<br>(3.40) | 8,148<br>(4.64) |
| Pondicherry | 54<br>(0.03) | 35<br>(0.02) | 44<br>(0.02) | 34<br>(0.02) | 33<br>(0.02) | 36<br>(0.02) |
| Andaman and Nicobar Islands | 34<br>(0.02) | 85<br>(0.05) | 38<br>(0.02) | 74<br>(0.04) | 46<br>(0.02) | 4<br>(0.01) |
| Lakshadweep | 3<br>(0.002) | 3<br>(0.002) | 3<br>(0.002) | 3<br>(0.002) | 3<br>(0.002) | 3<br>(0.002) |

*Source*: Government of Punjab, Statistical Abstract, Various Years.

*Note*: Percentages are shown in parentheses.

In Karnataka, the GCA has increased during pre as well as post-liberalisation periods except in 2000-01. It has increased from 10,660 thousand hectares in 1980-81 to 11,146 thousand hectares in 1985-86 and again increased to 12,284 thousand hectares in 2000-01 and further increased to 12,438 in 2006-07. The GCA has increased from 6,469 thousand hectares in 1980-81 to 6,632 thousand hectares in 1990-91 and further increased to 8,148 thousand hectares in 2006-07 in Tamil Nadu.

In Pondicherry, the GCA has declined by 35.19 per cent in 1985-86 as compared to 1980-81 and increased by 25.71 per cent in 1990-91 from 1985-86 and further increased by 9.09 per cent in 2006-07 from 2000-01. In Andaman and Nicobar Islands, it has increased by 150 per cent in 1985-86 from 1980-81 and declined by 55.29 per cent in 1990-91 from 1985-86 and again declined by 37.84 per cent in 2000-01 as compared to 1996-97 and further declined by 69.5 per cent in 2006-07 from 2000-01, whereas during pre as well as post-liberalisation periods 1980-81 to 2006-07, it is observed that in Lakshadweep, the GCA remains constant.

This table also indicates the percentage share of GCA in different states of south zone of India during 1980-81 to 2006-07. It is observed that in this zone, Andhra Pradesh has got first rank during 1980-81 to 2006-07. It has got 7.09 percentage share of GCA in 1980-81, 7.12, 7.26 and 7.29 in 1990-91, 2000-01 and 2006-07 respectively. Karnataka has got second rank followed by Tamil Nadu, Kerala during pre as well as post-liberalisation periods. During study, it is found that Pondicherry has got a little percentage share of GCA at country level. It is observed that in 1980-81, it has got 0.03 per cent of GCA and 0.02 per cent during 1985-86 to 2006-07, whereas in Andaman and Nicobar Islands a lot of variation is seen.

During post-liberalisation period (in 2006-07), the GCA in absolute terms has declined in all the states of south zone except in Karnataka, Tamil Nadu and Lakshadweep as compared to pre-liberalisation period (in 1990-91), whereas percentage share increased in all the states except in Pondicherry, Lakshadweep (percentage share remains constant) and in Andaman and Nicobar (percentage share declined). During pre-liberalisation period (1990-91), in Andhra Pradesh, GCA has increased approximately two times than that of Tamil Nadu, whereas Karnataka has got near about four times GCA as compared to Kerala, Pondicherry has received near about fifteen times as compared to Lakshadweep in 1990-91, on the other hand, Pondicherry has got twelve times more of GCA as compared to Lakshadweep in 2006-07.

The gross cropped area in the states of west zone of India during 1980-81 to 2006-07 is shown in Table 3.3. This table reveals that in pre-liberalisation period *i.e.* prior to 1990-91, Madhya Pradesh is the state, who has got the maximum *i.e.* 23,016 thousand hectares followed by Maharashtra and Rajasthan respectively. In 2006-07, the interesting fact is that it reversed from 1990-91 and Rajasthan is the state, who has got the maximum percentage share *i.e.* 20.67 followed by Gujarat, Madhya Pradesh, Maharashtra and Chhattisgarh.

This table indicates that in Madhya Pradesh, the GCA increased from 21,402 thousand hectares in 1980-81 to 23,880 thousand hectares in 1990-91 and further increased to 24,451 thousand hectares in 1996-97 and declined to 12,438 thousand hectares in 2006-07. Whereas in Maharashtra, the GCA has increased from 20,270 thousand hectares in 1980-81 to 21,855 in 1990-91 and declined to 8,148 thousand hectares in 2006-07. In Rajasthan, GCA has increased from 17,350 thousand hectares in 1980-81 to 19,380 thousand hectares in 1990-91 and further increased to 20,693 thousand hectares in 1996-97 and declined to 19,230 thousand hectares in 2000-01. The GCA has declined from 10,695 thousand hectares in 1980-81 to 9683 thousand hectares in 1985-86 and increased to 10,361 in 1990-91 and further increased to 12,811 thousand hectares in 2006-07 in Gujarat.

**Table 3.3: State-wise Distribution of Gross Cropped Area in West Zone in India during 1980-81 to 2006-07 (In 000 hectares)**

| *Years/States* | *West Zone* | | | | | |
|---|---|---|---|---|---|---|
| | *1980-81* | *1985-86* | *1990-91* | *1996-97* | *2000-01* | *2006-07* |
| Gujarat | 10,695<br>(6.17) | 9,683<br>(5.45) | 10,361<br>(5.59) | 11,001<br>(5.83) | 10,690<br>(5.73) | 12,811<br>(7.29) |
| Madhya Pradesh | 21,402<br>(12.35) | 23,016<br>(12.96) | 23,880<br>(12.88) | 24,451<br>(12.96) | 17,870<br>(9.58) | 12,438<br>(7.08) |
| Chhattisgarh | – | – | – | – | 5,327<br>(2.86) | 2,918<br>(1.66) |
| Maharashtra | 20,270<br>(11.69) | 20,537<br>(11.57) | 21,855<br>(11.79) | 21,722<br>(11.52) | 21,911<br>(11.74) | 8,148<br>(4.64) |
| Rajasthan | 17,350<br>(10.01) | 18,137<br>(10.22) | 19,380<br>(10.45) | 20,693<br>(10.97) | 19,230<br>(10.31) | 36,315<br>(20.67) |
| Goa and Daman and Diu | 141<br>(0.08) | 229<br>(0.13) | 158<br>(0.09) | 205<br>(0.11) | 178<br>(0.10) | 175<br>(0.10) |
| Dadra Nagar Haveli | 24<br>(0.01) | 26<br>(0.01) | 25<br>(0.01) | 25<br>(0.01) | 25<br>(0.01) | 28<br>(0.02) |

*Source*: Government of Punjab, Statistical Abstract, Various Years

*Note*: Percentages are shown in parentheses.

In Chhattisgarh, the GCA has declined by 7.08 per cent in 2006-07 as compared to 2000-01. In Goa and Daman and Diu, the GCA has increased by 62.41 per cent in 1985-86 as compared to 1985-86 and declined by 31 per cent in 1990-91 from 1985-86 and increased by 178 per cent in 1996-97 from 1990-91and declined by 13.17 per cent in 2000-01 from 1996-97 and further declined by 1.69 per cent in 2006-07 from 2000-01, whereas during 1990-91 to 1996-97, it is observed that in Dadra Nagar Haveli the GCA remains constant.

Percentage share analysis shows that in west zone, Madhya Pradesh has got first rank during 1980-81 to 1996-97 by receiving the maximum percentage share (12.35 per cent, 12.96 per cent and 12.96 per cent in 1980-81, 1985-86 and 1996-97 respectively), whereas it lost its first rank during 2000-01 to 2006-07. Maharashtra has got second rank during 1980-81 to 1996-97 and first in 2000-01. This state has got 11.69 per cent, 11.52 per cent and 4.64 per cent of GCA in 1980-81, 1996-97 and 2006-07 respectively. Rajasthan has got third position during 1980-81 to 1996-97, second in 2000-01 to 2006-07. This state is followed by Chhattisgarh, Goa and Daman and Diu and Dadra Nagar Haveli throughout the study period.

In all the states of west zone, a lot of variation is observed in absolute terms as well as in percentage share during the study period. As post-liberalisation period (2006-07) is compared to pre-liberalisation period (1990-91), in Gujarat the GCA has increased by 1.2 times and declined near about two times in Madhya Pradesh. In 1990-91, Madhya Pradesh has got 2.3 times more of GCA than that of Gujarat and

Maharashtra has received 1.13 times as compared to Rajasthan. Goa, Daman and Diu have got more than six times as compared to Dadra Nagar Haveli, whereas same pattern is also found in post-liberalisation period (2006-07).

The gross cropped area of different states of north zone of India during 1980-81 to 2006-07 is shown in Table 3.4. This table shows that Uttar Pradesh is ahead among all the other states by receiving maximum percentage share followed by Punjab and Haryana during the study period, only in 2006-07, Haryana superseded. It shows that in Haryana and Jammu & Kashmir, the GCA has increased during pre as well as post-liberalisation periods, whereas in Punjab as well as in Uttar Pradesh, it has declined in 2006-07, on the other hand, a lot of variation is seen in Himachal Pradesh.

**Table 3.4: State-wise Distribution of Gross Cropped Area in North Zone in India during 1980-81 to 2006-07 (In 000 hectares))**

| *Years/States* | *North Zone* | | | | | |
|---|---|---|---|---|---|---|
| | *1980-81* | *1985-86* | *1990-91* | *1996-97* | *2000-01* | *2006-07* |
| Haryana | 5,462 (3.15) | 5,601 (3.16) | 5,919 (3.19) | 6,074 (3.22) | 6,115 (3.28) | 6,394 (3.64) |
| Punjab | 6,763 (3.90) | 7,158 (4.03) | 7,502 (4.05) | 7,842 (4.16) | 7,941 (4.26) | 4,229 (2.41) |
| Uttar Pradesh | 24,574 (14.18) | 25,081 (14.13) | 25,480 (13.74) | 26,129 (13.85) | 27,057 (14.50) | 25,800 (14.69) |
| Jammu & Kashmir | 974 (0.56) | 1,030 (0.58) | 1,066 (0.57) | 1,077 (0.57) | 1,115 (0.60) | 1,126 (0.64) |
| Delhi | 87 (0.05) | 74 (0.04) | 71 (0.04) | 63 (0.03) | 57 (0.03) | 43 (0.02) |
| Himachal Pradesh | 946 (0.55) | 974 (0.55) | 931 (0.50) | 947 (0.50) | 948 (0.51) | 947 (0.54) |
| Uttarakhand | – | – | – | – | – | 1,241 (0.71) |

*Source*: Government of Punjab, Statistical Abstract, Various Years.

*Note*: Percentages are shown in parentheses.

In Uttar Pradesh, the GCA increased from 24,574 thousand hectares in 1980-81 to 25,081 thousand hectares in 1985-86 and further increased to 26,129 thousand hectares in 2000-01 and declined to 25,800 thousand hectares (4.65 per cent in 2006-07 as compared to 2000-01) in 2006-07. In Punjab, it has increased from 6,763 thousand hectares in 1980-81 to 7,502 thousand hectares in 1990-91 and further increased to 7,842 thousand hectares in 1996-97 and declined to 4,229 thousand hectares (46.74 per cent as compared to 2000-01), whereas in Haryana, it has increased by 2.54 per cent in 1985-86,5.68 per cent in 1990-91,2.6 per cent in 1996-97,0.68 per cent in 2000-01 and 4.56 per cent in 2006-07 as compared to their predecessor time period in the table. In Jammu & Kashmir, it has increased by 5.75 per cent in 1980-81, 3.5 per cent in 1985-86, 1.03 per cent in 1996-97, 3.53 per cent in 2000-01 and 0.99 per cent in 2006-07 from the years, which are predecessor in the table like 1985-86 is compared to 1980-81 and so on.

Proportionate-wise analysis gives a clear picture as it is found that Uttar Pradesh has got first rank during 1980-81 to 2006-07 among all the other states by getting maximum percentage share of GCA at country level. It has got 14.18 per cent of GCA in 1980-81, which has declined to 13.74 per cent in 1990-91 and increased to 14.50 per cent in 2000-01 and further increased to 14.69 per cent in 2006-07. Punjab has got second position during 1980-81 to 2000-01 and third in 2006-07 by getting 2.41 per cent at India level. The percentage share has increased from 3.90 in 1980-81 to 4.26 in 2000-01 and declined to 2.41 in 2006-07 in Punjab.

Haryana has got third rank during 1980-81 to 2000-01 and second in 2006-07. This state has got 3.15 per cent and 3.64 per cent in 1980-81 and 2006-07 respectively. Jammu & Kashmir has occupied fourth position, Himachal Pradesh fifth and Delhi sixth during pre as well as post-liberalisation periods.

From the above table, it is observed that GCA in absolute terms in all the states of north zone has increased during 1980-81 to 2000-01 and declined in Delhi, whereas a lot of variation is seen in Himachal Pradesh. In Haryana, Uttar Pradesh, Jammu Kashmir and in Himachal Pradesh, it has increased by 1.02 times in 2006-07 as compared to 1990-91. During pre-liberalisation period (1990-91), Punjab has received more than seven times as compared to Jammu & Kashmir and Himachal Pradesh has got thirteen times more from Delhi. Whereas during post-liberalisation period (2006-07), Punjab has got near about four times more from Jammu & Kashmir and Himachal Pradesh has received more than twenty two times than that of Delhi. It is found that Uttar Pradesh has got four times more of GCA as compared to Haryana during pre as well as post-liberalisation periods.

The gross cropped area of different states of east zone in India during 1980-81 to 2006-07 is shown in Table 3.5. This table indicates that Bihar is ahead among all the other states of east zones followed by West Bengal and Odisha throughout the study period. In Bihar, the GCA has declined from 11,148 thousand hectares in 1980-81 to 10,485 thousand hectares in 1990-91 and further declined to 7,582 thousand hectares (24.54 per cent as compared to 2000-01) in 2006-07. In West Bengal, it has increased from 7,620 thousand hectares in 1980-81 to 9,059 in 1996-97 and declined to 5,751 thousand hectares (36.92 per cent from 2000-01) in 2006-07. The GCA has increased from 8,746 thousand hectares in 1980-81 to 9,594 thousand hectares in 1990-91 and declined to 4,270 thousand hectares in 2006-07 in Odisha.

This table further reveals that in Bihar, the GCA has declined by 5.66 per cent in 1985-86 from 1980-81 and further declined by 24.54 per cent in 2006-07 from 2000-01. In Odisha, the increasing percentage of GCA has increased by 5.87 per cent in 1985-86 from 1980-81 and further increased by 3.62 per cent in 1990-91 from 1985-86 and declined by 14.36 per cent in 1996-97 from 1990-91 and further declined by 45.8 per cent in 2006-07 from 2000-01, whereas in West Bengal, the increasing rate has increased from 4.82 per cent in 1985-86,8.45 per cent in 1990-91,4.58 per cent in 1996-97 and 0.64 per cent in 2000-01 as compared to predecessor time periods in the given table. It is also seen that in the same state, the GCA declined by 36.92 per cent in 2006-07 from 2000-01.

**Table 3.5: State-wise Distribution of Gross Cropped Area in East Zone in India during 1980-81 to 2006-07 (In 000 Hectares)**

| *Years/States* | *East Zone* | | | | | |
|---|---|---|---|---|---|---|
| | *1980-81* | *1985-86* | *1990-91* | *1996-97* | *2000-01* | *2006-07* |
| Bihar | 11,148<br>(6.43) | 10,517<br>(5.92) | 10,485<br>(5.66) | 10,141<br>(5.38) | 10,048<br>(5.39) | 7,582<br>(4.32) |
| Jharkhand | – | – | – | – | – | 2,643<br>(1.50) |
| Odisha | 8,746<br>(5.05) | 9,259<br>(5.22) | 9,594<br>(5.17) | 8,216<br>(4.36) | 7,878<br>(4.22) | 4,270<br>(2.43) |
| West Bengal | 7,620<br>(4.40) | 7,987<br>(4.50) | 8,662<br>(4.67) | 9,059<br>(4.80) | 9,117<br>(4.89) | 5,751<br>(3.27) |

*Source*: Government of Punjab, Statistical Abstract, Various Years.

*Note*: Percentages are shown in parentheses.

It is found that Bihar has achieved topmost position in this zone among all the other states during the study. It has got 6.43 per cent in 1980-81 and 4.32 per cent in 2006-07. In Jharkhand, the GCA is 2643 thousand hectares in 2006-07, which is 1.5 per cent at all India level. The percentage share of Odisha has increased from 5.05 in 1980-81 to 5.22 in 1985-86 and declined to 4.22 in 2000-01 and again increased to 2.43 in 2006-07. West Bengal has occupied third rank during 1980-81 to 1990-91 and second during 1996-97 to 2006-07.

From the above table, it is found that during pre-liberalisation in Bihar state of east zone, GCA has declined in absolute terms, on the other hand, increased in Odisha as well as in West Bengal. During post-liberalisation period, it has declined in all the states. As the year 2006-07 is compared to the year 1990-91, it is observed that in Bihar, it has declined by eighteen times, in Odisha declined by 2.25 times and in West Bengal also declined by 1.51 times. Bihar has got 1.2 times more of GCA and 1.3 times as compared to West Bengal during pre as well post-liberalisation periods respectively.

The gross cropped area (GCA) of states in north-east zone of India during 1980-81 to 2006-07 is shown in Table 3.6. State-wise percentage share at the India level shows that Assam has occupied the first position by getting maximum percentage share of GCA followed by Tripura, Manipur, Meghalaya, Nagaland, Mizoram, Sikkim and Arunachal Pradesh during pre as well as post-liberalisation periods.

This table indicates that in Assam, the GCA has increased from 3,446 thousand hectares in 1980-81 to 3,797 thousand hectares in 1990-91 and further increased to 3,981 thousand hectares in 1996-97 and declined by 12.6 per cent in 2006-07 as compared to 2000-01. In Tripura, the GCA has increased from 375 thousand hectares in 1980-81 to 445 thousand hectares in 1990-91 and further increased to 458 thousand hectares in 1996-97 and declined to 310 thousand hectares in 2006-07, whereas the GCA has declined from 219 thousand hectares in 1980-81 to 183 thousand hectares in 1985-86 and increased to 231 thousand hectares in 2006-07 in Manipur. The same picture is seen in states Meghalaya, Nagaland, Mizoram and in Arunachal Pradesh.

**Table 3.6: State-wise Distribution of Gross Cropped Area in North-East Zone in India during 1980-81 to 2006-07 (In 000 Hectares)**

| Years/States | North-East Zone | | | | | |
|---|---|---|---|---|---|---|
| | 1980-81 | 1985-86 | 1990-91 | 1996-97 | 2000-01 | 2006-07 |
| Assam | 3,446 (1.99) | 3,794 (2.14) | 3,797 (2.05) | 3,981 (2.11) | 4,065 (2.18) | 3,553 (2.02) |
| Tripura | 375 (0.22) | 423 (0.24) | 445 (0.24) | 458 (0.24) | 428 (0.23) | 310 (0.18) |
| Manipur | 219 (0.13) | 183 (0.10) | 180 (0.10) | 203 (0.11) | 209 (0.11) | 231 (0.13) |
| Meghalaya | 223 (0.13) | 212 (0.12) | 243 (0.13) | 244 (0.13) | 314 (0.17) | 1057 (0.60) |
| Nagaland | 158 (0.09) | 197 (0.11) | 210 (0.11) | 246 (0.13) | 288 (0.15) | 657 (0.37) |
| Mizoram | 106 (0.06) | 71 (0.04) | 74 (0.04) | 109 (0.06) | 94 (0.05) | 213 (0.12) |
| Sikkim | 91 (0.05) | 134 (0.08) | 152 (0.08) | 142 (0.08) | 126 (0.07) | 155 (0.09) |
| Arunachal Pradesh | 141 (0.08) | 149 (0.08) | 245 (0.13) | 240 (0.13) | 263 (0.14) | 275 (0.16) |

*Source*: Government of Punjab, Statistical Abstract, Various Years.

*Note*: Percentages are shown in parentheses.

This table reveals that Assam has received 1.99 per cent of GCA at national level in 1980-81, increased to 2.14 per cent in 1985-86 and declined to 2.05 in 1990-91 and again increased to 2.18 per cent in 2000-01, which has again declined to 2.02 per cent in 2006-07. Tripura has received 0.22 per cent in 1980-81, 0.24 per cent, 0.23 per cent and 0.18 per cent in 1985-86, 2000-01 and 2006-07 respectively.

It is found that Manipur has got 0.13 per cent in 1980-81 and in 2006-07, in mid years there is declined in percentage share of GCA. Meghalaya has got 0.13 percentage share in 1980-81, which has increased to 0.60 in 2006-07. During study it is observed that percentage share of Nagaland is increased, whereas during 1985-86 to 1990-91, the percentage share of GCA remains constant.

In Mizoram, the percentage share *i.e.* 0.04 per cent remains constant in 1985-86 and 1990-91, 0.06 per cent in both years of 1980-81, 1996-97 and 0.12 per cent in 2006-07. Sikkim has got 0.05 per cent and 0.09 per cent in 1980-81 and 2006-07 respectively. Arunachal Pradesh received 0.08 per cent and 0.16 per cent in 1980-81and 2006-07 respectively.

It is observed that in all the states of north-east zone, the GCA in absolute terms has increased in Nagaland state only during pre as well as post-liberalisation periods, whereas a lot of variation is seen in percentage share. As the year 2006-07 is compared to the year 1990-91, it is observed that in Manipur, the GCA has increased by 1.28 times, in Sikkim 1.02 times and in Arunachal Pradesh 1.12 times, whereas in Meghalaya, it has risen up by 4.35 times and in Nagaland more than three times. During 1990-91, Assam has got more than eight times as compared to Tripura, Meghalaya more than three times and Arunachal Pradesh near about two times more of GCA than that of Sikkim (same pattern is also found during post-liberalisation period in Arunachal Pradesh).

## Chapter 4

# Agricultural Subsidies in India

Indian government play vital role in the development of agriculture sector. This vital role can take a number of farms like minimum support price, direct payment, input subsidies etc. to influence the cost and availability of farm inputs. Agriculture subsidies are a governmental financial support given to the farmers to supplement their income. In this chapter different types of subsidies are analysed.

## Section I

The total subsidies in India during 1980-81 to 2008-09 are shown in Table 4.1. In 1980-81, the total subsidies have increased from Rs.1,228.54 crores to Rs.4,796.16 crores in 1985-86 and further increased to Rs.1,15,952.20 crores in 2008-09. The fertilizers subsidies have increased from Rs.471 crores in 1980-81 to Rs.13,724.05 crores in 2000-01 and further increased to Rs.1,01,180.68 crores in 2008-08, whereas the electricity subsidy has increased from Rs. 357.56 crores in 1980-81 to Rs.4,621 crores in 1990-91 and further increased to Rs.26,904 crores in 2000-01 and declined to Rs.14,771.52 crores in 2008-09. The irrigation subsidy (Canal water) has increased from Rs. 399.10 crores in 1980-81 to Rs.3,917.41 crores in 1990-91 and further increased to Rs.14,711.71 crores in post-liberalisation period (2000-01).

The percentage share of fertilizers subsidies in total subsidies has declined from 38.41 in 1980-81 to 35.20 in 1990-91 and further declined to 24.80 in 2000-01 and increased to 87.26 in 2008-09. Whereas the percentage share of electricity subsidy has increased from 29.10 in 1980-81 to 35.07 in 1990-91 and further increased to 48.62 in 2000-01 and declined to 12.74 in 2008-09, on the other hand, the percentage share of irrigation subsidy is 32.49, 34.76 and 26.58 in 1980-81, 1985-86 and 2000-01 respectively.

From the Table 4.1, it is found that at national level during pre as well as post-liberalisation periods, the total subsidies have increased at different increasing rates and in absolute terms. In 2008-09, the total subsidies have increased by 94.38 times than that of 1980-81, whereas fertilizers subsidies twenty nine times, electricity subsidy 75.24 times and irrigation subsidy by 36.86 times in 2000-01 as compared to pre-liberalisation period (1980-81). In pre-liberalisation period (1990-91), fertilizers subsidies have increased by 1.00 times and 1.18 times than that of electricity subsidy and irrigation subsidy respectively. In 2000-01, electricity subsidy has increased by 1.83 times more and 1.96 times than that of irrigation subsidy and fertilizers subsidy.

**Table 4.1: Distribution of Total Subsidies in India during 1980-81 to 2008-09 (In Rs. Crores)**

| *Subsidies/Years* | *Fertilizers* | *Electricity* | *Irrigation* | *Total* |
|---|---|---|---|---|
| 1980-81 | 471.88<br>(38.41) | 357.56<br>(29.10) | 399.10<br>(32.49) | 1,228.54<br>(100.00) |
| 1985-86 | 1,804.80<br>(37.63) | 1,324.15<br>(27.61) | 1,667.21<br>(34.76) | 4,796.16<br>(100.00) |
| 1990-91 | 4,638.56<br>(35.20) | 4,621.00<br>(35.07) | 3,917.41<br>(29.73) | 13,176.97<br>(100.00) |
| 1996-97 | 8,148.41<br>(23.86) | 15,594.00<br>(45.67) | 10,404.73<br>(30.47) | 34,147.14<br>(100.00) |
| 2000-01 | 13,724.05<br>(24.80) | 26,904.00<br>(48.62) | 14,711.71<br>(26.58) | 55,339.76<br>(100.00) |
| 2008-09 | 1,01,180.68<br>(87.26) | 14,771.52<br>(12.74) | –<br>– | 1,15,952.20<br>(100.00) |

*Source*: (1) Government of India, Fertilizers Association, Fertilizer Statistics, various issues, New Delhi.

(2) Government of India, State Electricity Boards, Annual Reports, Various Years.

*Note*: Percentages are shown in parentheses.

Zone-wise distribution of total subsidies in India during 1980-81 to 2008-09 is shown in Table 4.2. This table shows that in all the zones of India, total subsidies have increased during pre as well as post-liberalisation periods. In south zone, these have increased from Rs.354.61 crores in 1980-81 to Rs.3,397.63 crores in 1990-91 and further increased to Rs.30,300.42 crores in 2008-09, whereas in west zone, these have increased from Rs.311.23 crores in 1980-81 to Rs.4,430.34 crores in 1990-91 and further increased to Rs.32,581.17 crores in 2008-09. In north zone, total subsidies have risen up from Rs.448.29 crores in 1980-81 to Rs.3,985.42 crores in 1990-91 and further risen up to Rs.36,852.04 crores in 2008-09. On the other hand, in east zone, these have increased from Rs.103.58 crores in 1980-81 to Rs.1,261.16 crores in 1990-91 and further increased to Rs.15,174.20 crores in 2008-09 and north-east zone, has got Rs.10.84 crores, Rs.36.57 crores and to Rs.1,044.36 crores in 1980-81, 1985-86 and 2008-09 respectively.

Percentage-wise analysis shows that the north zone has got topmost position by receiving (36.49 per cent in 1980-81 and 32.20 per cent in 1985-86) during pre-liberalisation period, whereas west zone has got first position (33.62 per cent in

1990-91 and 41.47 per cent in 1996-97) during first phase of liberalisation period. West zone is leading among all the other states by getting the maximum percentage share (39.65 per cent) followed by south zone, north zone, east zone, and north-east zone in 2000-01, whereas in 2008-09 north zone is ahead among all the other states by receiving major the percentage share (31.80 per cent), followed by west zone, south zone, east zone and north-east zone.

**Table 4.2: Zone-wise Distribution of Total Subsidies in India during 1980-81 to 2008-09 (In Rs. Crores)**

| *Years/States* | *1980-81* | *1985-86* | *1990-91* | *1996-97* | *2000-01* | *2008-09* |
|---|---|---|---|---|---|---|
| South | 354.61<br>(28.86) | 1,292.88<br>(26.96) | 3,397.63<br>(27.78) | 9,691.15<br>(28.38) | 16,693.26<br>(30.18) | 30,300.42<br>(26.13) |
| West | 311.23<br>(25.33) | 1,413.15<br>(29.46) | 4,430.34<br>(33.62) | 14,162.44<br>(41.47) | 21,931.10<br>(39.65) | 32,581.17<br>(28.09) |
| North | 448.29<br>(36.49) | 1,544.41<br>(32.20) | 3,985.42<br>(30.25) | 7,634.06<br>(22.36) | 12,200.55<br>(22.01) | 36,852.04<br>(31.80) |
| East | 103.58<br>(8.43) | 509.14<br>(10.62) | 1,261.16<br>(9.57) | 2,534.88<br>(7.42) | 4,247.28<br>(7.68) | 15,174.20<br>(13.09) |
| North-East | 10.84<br>(0.88) | 36.57<br>(0.76) | 102.42<br>(0.78) | 124.62<br>(0.36) | 267.58<br>(0.48) | 1,044.36<br>(0.90) |
| India | 1,228.54<br>(100) | 4,796.16<br>(100) | 13,176.97<br>(100) | 34,147.14<br>(100) | 55,316.13<br>(100) | 1,15,977.84<br>(100) |

*Source*: (1) Government of India, Fertilizers Association, Fertilizer Statistics, various issues, New Delhi.

(2) Government of India, State Electricity Boards, Annual Reports, Various Years.

*Note*: (1) Total subsidies are calculated by adding subsidies of electricity, irrigation and fertilizers only given in zone basis.

(2) Percentages are shown in parentheses.

From the above table, it is concluded that total subsidies at national level as well as zone level have increased in absolute terms during pre and post-liberalisation periods. At national level in 2008-09, these have risen up more than nine times as compared to 1990-91. As zone-wise, in east zone these have increased the maximum *i.e.* twelve times among all the other zones, whereas in west zone these have increased by only seven times.

The state-wise distribution of total subsidies in south zone in India during 1980-81 to 2008-09 is shown in Table 4.3. This table indicates that Andhra Pradesh is ahead among all the other states in 1980-81, whereas it has lost its topmost position *i.e.* Tamil Nadu during 1985-86 to 2008-09. In Andhra Pradesh, the total subsidies have increased from Rs.126.14 crores in 1980-81 to Rs.134.08 crores in 1990-91 and further increased to Rs.140.22 crores in 2008-09. In Tamil Nadu, these have increased from Rs.132.89 crores in 1980-81 to Rs.902.22 crores in 1990-91 and further increased to Rs.6,970.83 crores in 2008-09.

In Karnataka, these have risen up from Rs.79.51 crores in 1980-81 to Rs.858.96 crores in 1990-91 and declined to Rs.100.40 crores in 2008-09, whereas in Kerala, these have declined from Rs.16.06 crores in 1980-81 to Rs.9.79 crores in 2000-01 and increased to Rs.20.34 crores in 2008-09. In Pondicherry, these have risen up from Rs.11.96 crores in 1996-97 to Rs.113.49 crores in 2008-09, whereas these have increased from Rs.0.22 crores in 1996-97 to Rs.2.44 crores in 2008-09 in Andaman and Nicobar Islands.

Percentage-wise analysis indicates that Andhra Pradesh has got topmost position by getting the maximum percentage share during pre as well as post-liberalisation period. Its percentage share has increased from 10.27 in 1980-81 to 11.04 in 1990-91 and further increased to 13.79 in 2000-01, whereas the percentage share of Karnataka has increased from 6.47 in 1980-81 to 8.52 in 2000-01 and declined to 7.70 in 2008-09.

The percentage share of Kerala has declined from 1.31 in 1980-81 to 0.92 in 1996-97 and increased to 1.56 in 2008-09, on the other hand, the percentage share in Tamil Nadu has declined from 10.82 in 1980-81 to 6.01 in 2008-09. During 1996-97 to 2000-01, the percentage share of Pondicherry as well as in Andaman and Nicobar Islands remains constant.

**Table 4.3: State-wise Distribution of Total Subsidies in South Zone in India during 1980-81 to 2008-09 (In Rs. Crores)**

| *Years/States* | *South Zone* | | | | | |
|---|---|---|---|---|---|---|
| | *1980-81* | *1985-86* | *1990-91* | *1996-97* | *2000-01* | *2008-09* |
| Andhra Pradesh | 126.14<br>(10.27) | 499.05<br>(10.41) | 1,454.12<br>(11.04) | 3,783.90<br>(11.08) | 7,671.34<br>(13.79) | 12,473.73<br>(10.76) |
| Karnataka | 79.51<br>(6.47) | 361.09<br>(7.53) | 858.96<br>(6.52) | 2,928.34<br>(8.58) | 4,738.68<br>(8.52) | 8,930.97<br>(7.70) |
| Kerala | 16.06<br>(1.31) | 53.29<br>(1.11) | 181.99<br>(1.38) | 313.76<br>(0.92) | 419.18<br>(0.76) | 1,808.96<br>(1.56) |
| Tamil Nadu | 132.89<br>(10.82) | 379.45<br>(7.91) | 902.55<br>(6.85) | 1831.79<br>(5.36) | 3,843.08<br>(6.91) | 6,970.83<br>(6.01) |
| Pondicherry | – | – | – | 11.96<br>(0.04) | 20.64<br>(0.04) | 113.49<br>(0.10) |
| Andaman and Nicobar Islands | – | – | – | 0.22<br>(0.001) | 0.35<br>(0.001) | 2.44<br>(0.002) |
| Lakshadweep | – | – | – | 821.18<br>(2.40) | – | – |

*Source*: (1) Government of India, Fertilizers Association, Fertilizer Statistics, various issues, New Delhi.
(2) Government of India, State Electricity Boards, Annual Reports, Various Years.

*Note*: (1) Total subsidies are calculated by adding subsidies of electricity, irrigation and fertilizers.
(2) Percentages are shown in parentheses.

During pre as well as post-liberalisation periods, the total subsidies have increased in absolute terms in all the states of south zone. As compared to post-liberalisation period (2008-09) to pre-liberalisation period (1990-91), it is found that in Karnataka, these have increased by more than ten times, in Kerala near about ten times, in Andhra Pradesh near about nine times and in Tamil Nadu near about eight times. Andhra Pradesh has got near about two times more of total subsidies as compared to Karnataka during pre as well post-liberalisation periods.

The state-wise distribution of total subsidies in west zone in India during 1980-81 to 2008-09 is shown in Table 4.4. This table shows that in all the states (except in Goa) of this zone, the total subsidies have increased during pre as well as post-liberalisation periods, except in 2000-01. In Gujarat, these have increased from Rs.80.21 crores in 1980-81 to Rs.7,756.36 in 2000-01 and further increased to Rs.8,074.23 crores in 2008-09, whereas Madhya Pradesh has received Rs. 27.19 crores, Rs.2,747.87 crores, Rs.4,318.16 crores and Rs.8,074.23 crores in 1980-81, 1996-97, 2000-01 and 2008-09 respectively.

**Table 4.4: State-wise Distribution of Total Subsidies in West Zone in India during 1980-81 to 2008-09 (In Rs. Crores)**

| *Years/States* | *West Zone* | | | | | |
|---|---|---|---|---|---|---|
| | *1980-81* | *1985-86* | *1990-91* | *1996-97* | *2000-01* | *2008-09* |
| Gujarat | 80.21<br>(6.53) | 343.78<br>(7.17) | 1,200.63<br>(9.11) | 4,034.88<br>(11.82) | 7,756.36<br>(13.95) | 8,074.23<br>(6.96) |
| Madhya Pradesh | 27.19<br>(2.21) | 180.32<br>(3.76) | 751.87<br>(5.71) | 2,747.87<br>(8.05) | 4,318.16<br>(7.76) | 6,688.07<br>(5.77) |
| Chhattisgarh | – | – | – | – | – | 1931.44<br>(1.67) |
| Maharashtra | 138.50<br>(11.27) | 607.56<br>(12.67) | 1,920.84<br>(14.58) | 5,376.38<br>(15.74) | 6,351.63<br>(11.42) | 10,423.32<br>(8.99) |
| Rajasthan | 99.70<br>(8.12) | 377.41<br>(7.87) | 1,190.06<br>(9.03) | 3,460.62<br>(10.13) | 5,014.51<br>(9.02) | 5,476.55<br>(4.72) |
| Goa | 0.20<br>(0.02) | 1.51<br>(0.03) | 2.90<br>(0.02) | 28.58<br>(0.08) | 50.70<br>(0.09) | 33.02<br>(0.03) |
| Daman and Diu | – | – | – | 0.185<br>(0.001) | 0.214<br>(0.0004) | 1.584<br>(0.001) |
| Dadra Nagar Haveli | – | – | – | 0.575<br>(0.002) | 0.758<br>(0.001) | 4.468<br>(0.004) |

*Source*: (1) Government of India, Fertilizers Association, Fertilizer Statistics, various issues, New Delhi.
(2) Government of India, State Electricity Boards, Annual Reports, Various Years.

*Note*: (1) Total subsidies are calculated by adding subsidies of electricity, irrigation and fertilizers.
(2) Percentages are shown in parentheses.

In Maharashtra, total subsidies have increased from Rs.138.50 crores in 1980-81 to Rs.6,351.63 crores in 2000-01 and further increased to Rs.20,423.32 crores in 2008-09. Rajasthan has got Rs.99.70 crores, Rs.1,190.06 crores, Rs.3,460.62 crores and Rs.5,476.55 crores in 1980-81, 1990-91, 1996-97 and 2008-09 respectively. These subsidies have increased during pre as well as post-liberalisation periods except in 2008-09 in Goa.

Percentage-wise analysis reveals that Maharashtra is leading all the other states of west zone by getting major percentage share of total subsidies during pre as well as post-liberalisation periods. Madhya Pradesh has got 2.21 per cent, 8.05 per cent and 5.77 per cent, whereas Gujarat has got 6.53 per cent, 11.82 per cent and 6.96 per cent in 1980-81, 1996-97, and 2008-09 respectively. The percentage share of Goa has increased from 0.02 per cent in 1980-81 to 0.09 in 2000-01 and declined to 0.03 in 2008-09, whereas Daman and Diu and Dadra Nagar Haveli have received a little percentage share of total subsidies during pre as well as post-liberalisation periods.

In all the states of west zone, the total subsidies including fertilizers, electricity and irrigation subsidies have increased in absolute terms, whereas percentage-wise analysis shows a lot of variation during the study period. During pre-liberalisation period (1990-91), Gujarat as well as Maharashtra has got near about two times more of total subsidies as compared to Madhya Pradesh and Rajasthan respectively, whereas during post-liberalisation period (2008-09), Gujarat has got 1.2 times more of total subsidies than that of Madhya Pradesh and Maharashtra has got 1.2 times more of total subsidies as compared to Rajasthan.

The state-wise distribution of total subsidies in north zone in India during 1980-81 to 2008-09 is shown in Table 4.5. This table shows that in all the states (except in Jammu & Kashmir and in Delhi) of north zone, total subsidies have increased during pre as well as post-liberalisation periods. Uttar Pradesh is followed by Punjab, Haryana, Jammu & Kashmir Himachal Pradesh and Delhi during pre as well as post-liberalisation periods.

**Table 4.5: State-wise Distribution of Total Subsidies in North Zone in India during 1980-81 to 2008-09 (In Rs. Crores)**

| *Years/States* | *North Zone* | | | | | |
|---|---|---|---|---|---|---|
| | *1980-81* | *1985-86* | *1990-91* | *1996-97* | *2000-01* | *2008-09* |
| Haryana | 77.78<br>(6.33) | 219.29<br>(4.57) | 575.27<br>(4.37) | 1,533.81<br>(4.49) | 3,142.21<br>(5.65) | 7,872.80<br>(6.79) |
| Punjab | 116.75<br>(9.50) | 435.15<br>(9.07) | 1146.23<br>(8.70) | 1,795.13<br>(5.26) | 3,798.07<br>(6.83) | 9,783.20<br>(8.44) |
| Uttar Pradesh | 246.64<br>(20.08) | 855.90<br>(17.75) | 2,173.66<br>(16.50) | 4,132.43<br>(12.10) | 5,061.87<br>(9.10) | 17,912.73<br>(15.44) |
| Jammu & Kashmir | 4.72<br>(0.38) | 23.35<br>(0.49) | 65.89<br>(0.58) | 136.03<br>(0.40) | 122.56<br>(0.22) | 426.91<br>(0.37) |
| Delhi | – | – | – | 12.17<br>(0.04) | 4.21<br>(0.01) | 2.60<br>(0.002) |
| Himachal Pradesh | 2.40<br>(0.20) | 10.72<br>(0.22) | 24.37<br>(0.18) | 24.45<br>(0.07) | 47.99<br>(0.09) | 233.03<br>(0.20) |
| Uttarakhand | – | – | – | – | – | 646.43<br>(0.56) |

*Source*: (1) Government of India, Fertilizers Association, Fertilizer Statistics, various issues, New Delhi.

(2) Government of India, State Electricity Boards, Annual Reports, Various Years.

*Note*: (1) Total subsidies are calculated by adding subsidies of electricity, irrigation and fertilizers.

(2) Percentages are shown in parentheses.

Uttar Pradesh has got Rs.246.64 crores, Rs.1423.70 crores, Rs.5,061.87 crores and Rs.17,912.73 crores, whereas Punjab these have increased from Rs.116.75 crores in 1980-81 to Rs.3,798.07 crores in 2000-01 and further increased to Rs.9,783.20 crores in 2008-09. In Haryana, these have increased from Rs.77.78 crores in 1980-81 to Rs.575.27 crores in 1990-91 and further increased to Rs.7,872.80 crores in 2008-09, whereas in Himachal Pradesh has received Rs.2.40 crores, Rs.24.37 crores, Rs.47.99 crores and Rs.233.03 crores in 1980-81, 1990-91, 2000-01 and 2008-09 respectively.

Percentages-wise analysis reveals that Uttar Pradesh has got topmost position by receiving major percentage share of total subsidies *i.e.* 20.08 per cent, 17.85 per cent, 16.50 per cent, 12.10 per cent, 9.10 per cent and 15.44 per cent in 1980-81, 1985-86, 1990-91, 1996-97, 2000-01 and 2008-09 respectively. The percentage share of Punjab has declined from 9.50 in 1980-81 to 5.26 in 1996-97 and increased to 8.44 in 2008-09. In Haryana, the percentage share has declined from 6.33 in 1980-81 to 5.65 in 2000-01 and further increased to 6.79 in 2008-09, whereas percentage share in Jammu & Kashmir has increased from 0.38 in 1980-81 to 0.58 in 1990-91 and declined to 0.22 in 2000-01 and again increased to 0.37 in 2008-09. It is observed that in Delhi the percentage share has declined during pre as well as post-liberalisation periods, whereas it has increased during pre as well as post-liberalisation periods except in Himachal Pradesh during 1990-91 to 1996-97.

**Table 4.6: State-wise Distribution of Total Subsidies in East Zone in India during 1980-81 to 2008-09 (In Rs. Crores)**

| *Years/States* | *East Zone* | | | | | |
|---|---|---|---|---|---|---|
| | *1980-81* | *1985-86* | *1990-91* | *1996-97* | *2000-01* | *2008-09* |
| Bihar | 42.37 (3.45) | 275.79 (5.75) | 638.55 (4.85) | 1,031.16 (3.02) | 1,881.41 (3.38) | 6,232.02 (5.37) |
| Jharkhand | – | – | – | – | – | 651.97 (0.56) |
| Odisha | 17.16 (1.40) | 55.86 (1.16) | 141.01 (1.07) | 566.84 (1.66) | 732.00 (1.32) | 2,172.72 (1.87) |
| West Bengal | 44.05 (3.59) | 177.49 (3.70) | 481.61 (3.65) | 936.87 (2.74) | 1,190.10 (2.14) | 6,171.27 (5.32) |

*Source*: (1) Government of India, Fertilizers Association, Fertilizer Statistics, various issues, New Delhi.

(2) Government of India, State Electricity Boards, Annual Reports, Various Years.

*Note*: (1) Total subsidies are calculated by adding subsidies of electricity, irrigation and fertilizers.

(2) Percentages are shown in parentheses.

From above, it is found that during the study period, the total subsidies have increased in absolute terms in all the states of north zone except in Jammu & Kashmir and Delhi. As compared to the year 2008-09 to the year 1990-91, it is observed that Haryana is that state where total subsidies have increased maximum number of times *i.e.* 13.6, whereas only 6.5 times in Jammu & Kashmir. As compared to Haryana, Punjab has got two times more of total subsidies in 1990-91 and 1.2 times in 2008-09, whereas Uttar Pradesh has received more than twenty one times more of total subsidies during pre-liberalisation period (1990-91), whereas during post-liberalisation period (2008-09), it has got near about fifty times more of total subsidies as compared to Jammu & Kashmir.

State-wise distribution of total subsidies in east zone in India during 1980-81 to 2008-09 is shown in Table 4.6. This table reveals that in all the state of east zone, the total subsidies have risen up during pre as well as post-liberalisation periods. West Bengal is leading the other states in 1980-81 followed by Bihar, Odisha, whereas during other years of study Bihar is leading the other states. In West Bengal, total subsidies have increased from Rs.44.05 crores in 1980-81 to Rs.6,171.27 crores in 2008-09. In Bihar, these have increased from Rs.42.37 crores in 1980-81 to Rs.6,232.02 crores in 2008-09, whereas these have increased from Rs.17.16 crores in 1980-81 to Rs.2,172.72 crores in 2008-09 in Odisha.

It is observed that Bihar has got maximum percentage share of total subsidies during pre as well as post-liberalisation periods except in 1980-81 as it has got 5.75 per cent, 3.38 per cent and 5.37 per cent in 1985-86, 2000-01 and 2008-09 respectively, In Odisha, the percentage share has increased from 1.40 in 1980-81 to 1.87 in 2008-09, whereas the percentage share has declined from 3.59 to 2.74 in 1996-97 and increased to 5.32 in 2008-09 in West Bengal.

Total subsidies have increased in absolute terms in all the states of east zone during pre as well post-liberalisation periods, whereas a lot of variation is seen in percentage share. As post-liberalisation period (2008-09) is compared to pre-liberalisation period (1990-91), in Odisha, these have risen up by 15.41 times more of total subsidies, West Bengal 12.8 times and near about ten times in Bihar. In 1990-91, Bihar has got 4.53 times more of total subsidies than that of Odisha.

The state-wise distribution of total subsidies in north-east zone in India during 1980-81 to 2008-09 is shown in Table 4.7. This table reveals that during 1980-81 to 1990-91 and in 2008-09, the total subsidies have increased in all the states, whereas during 1996-97 to 2000-01 these have declined in Tripura, Manipur, Mizoram and Sikkim.

In Assam, these have increased from Rs.8.81 crores in 1980-81 to Rs.896.10 crores in 2008-09, whereas in Tripura, these have increased from Rs.0.33 crores in 1980-81 to 8.55 crores in 1996-97 and further increased to 56.34 crores in 2008-09. In Manipur, these have increased from Rs.1.07 crores in 1980-81 to Rs.47.26 crores in 1996-97, it has declined to Rs.39.89 crores in 2000-01. Total subsidies have increased from Rs.0.23 crores in 1980-81 to Rs.0.63 crores in 1985-86 and further increased to Rs.14.99 crores in 2008-09 in Meghalaya.

Nagaland has got Rs.0.01 crores and Rs.3.70 crores in 1980-81 and 2008-09 respectively, whereas in Mizoram, total subsidies have increased from Rs.0.01 crores in 1980-81 to Rs.17.67 crores in 2008-09, whereas in Sikkim, these have increased from Rs.0.37 crores in 1980-81 to Rs.0.57 crores in 1990-91 and declined to Rs.0.31 crores in 2000-01. Arunachal Pradesh has got Rs.0.01 crores, Rs.0.16 crores, Rs.1.00 crores and Rs.3.29 crores in 1980-81, 1990-91, 2000-01 and 2008-09 respectively.

Percentage-wise analysis shows that in Assam, the percentage share has declined from 0.72 in 1980-81 to 0.19 in 1996-97 and increased to 0.77 in 2008-09, whereas the percentage share of Tripura in total subsides have increased throughout the study period except in 1996-97. In Manipur, the percentage share remains constant during 1980-81 to 1985-86, increased to 0.15 in 1990-91 and declined to 0.05 in 2008-09, whereas in Meghalaya, the percentage share remains constant during pre as well as post-liberalisation period. The percentage share of Mizoram has increased from 0.0005 in 1980-81 to 0.023 in 1990-91 and declined to 0.015 in 2008-09, Arunachal Pradesh has got 0.0004 per cent, 0.001 per cent, 0.002 per cent and 0.003 per cent in 1980-81, 1985-86, 1996-97 and 2008-09 respectively.

**Table 4.7: State-wise Distribution of Total Subsidies in North-East Zone in India during 1980-81 to 2008-09 (In Rs. Crores)**

| *Years/States* | *North-East Zone* | | | | | |
|---|---|---|---|---|---|---|
| | *1980-81* | *1985-86* | *1990-91* | *1996-97* | *2000-01* | *2008-09* |
| Assam | 8.81<br>(0.72) | 26.24<br>(0.55) | 63.03<br>(0.48) | 63.65<br>(0.19) | 256.92<br>(0.46) | 896.10<br>(0.77) |
| Tripura | 0.33<br>(0.03) | 4.42<br>(0.09) | 14.72<br>(0.11) | 8.55<br>(0.03) | 20.51<br>(0.04) | 56.34<br>(0.05) |
| Manipur | 1.07<br>(0.09) | 4.11<br>(0.09) | 19.47<br>(0.15) | 47.26<br>(0.14) | 39.89<br>(0.07) | 52.28<br>(0.05) |
| Meghalaya | 0.23<br>(0.02) | 0.63<br>(0.01) | 0.99<br>(0.01) | 3.37<br>(0.01) | 6.80<br>(0.01) | 14.99<br>(0.01) |
| Nagaland | 0.01<br>(0.001) | 0.05<br>(0.001) | 0.41<br>(0.003) | 0.44<br>(0.001) | 0.54<br>(0.001) | 3.70<br>(0.003) |
| Mizoram | 0.01<br>(0.0005) | 0.83<br>(0.017) | 3.08<br>(0.023) | 0.20<br>(0.0006) | 2.01<br>(0.004) | 17.67<br>(0.015) |
| Sikkim | 0.37<br>(0.03) | 0.25<br>(0.01) | 0.57<br>(0.004) | 0.38<br>(0.001) | 0.31<br>(0.001) | – |
| Arunachal Pradesh | 0.01<br>(0.0004) | 0.04<br>(0.001) | 0.16<br>(0.001) | 0.76<br>(0.002) | 1.00<br>(0.002) | 3.29<br>(0.003) |

*Source*: (1) Government of India, Fertilizers Association, Fertilizer Statistics, various issues, New Delhi.

(2) Government of India, State Electricity Boards, Annual Reports, Various Years.

*Note*: (1) Total subsidies are calculated by adding subsidies of electricity, irrigation and fertilizers.

(2) Percentages are shown in parentheses.

In absolute terms, the total subsidies have increased in all the states of north-east zone except in Tripura, Manipur, Mizoram and Sikkim throughout the study period. As the year 2008-09 is compared to the year 1990-91, it is observed that in Meghalaya, these have increased by 15.14 times more of total subsidies, whereas in Manipur, only near about three times. Assam has received 4.28 times in 1990-91 and near about sixteen times of total subsidies in 2008-09 as compared to Tripura, whereas Manipur has got 6.32 times more of total subsidies in 1990-91 and approximately three times in 2008-09 than that of Mizoram.

## Section – II

The Indian fertilizer industry has come a long way since its early days of post-independence era. Today, India is one of the largest producers and consumers of fertilizers in the world. The mounting burden of subsidies compelled the policy planners to make a serious attempt to reform the fertilizer price policy to rationalize the fertilizer subsidy. The Indian fertilizer industry, given its strategic importance in achieving self-sufficiency of food grains production in the coun-try, has for decades, been under government control. With the objective of providing fertilizers to farmers at an affordable price and ensuring adequate returns on investments to entrepreneurs, a fertilizer policy of providing fertilizers to farmers at subsidized prices was envisaged to induce farmers to use fertilizers (Sharma, 2010).

The subsidy was given for the first time in 1973-74 on the imported fertilizers on account of their steeply rising cost in awake of world oil crisis. Since then fertilizer subsidy increased from 0.07 per cent of G.D.P. in 1973-74 to (1.17 per cent of G.D.P. in 1989-90. Fertilizer is the vital input needed for increasing agricultural production. It has been the endeavour of the government to ensure availability of adequate quantity of fertilizer at reasonable prices to the farmers across the country over the years (Chander, 2000).

Zone-wise fertilizers subsidies in India during 1980-81 to 2008-09 are shown in Table 4.8. This table reveals that in India, these have increased from Rs.471.88 crores in 1980-81 to Rs.8,148.41 crores in 1996-97 and further increased to Rs.1,01,180.68 crores in 2008-09. North zone occupied first position during 1980-81 to 1990-91 and second in 1996-97 and first during 2000-01 to 2008-09. South zone has got second rank during 1980-81 to 1990-91 and first in 1996-97 and again occupied second in 2000-01 and third during 2005-06 to 2008-09.

West zone has ranked third during 1980-81 to 2000-01 and second during 2005-06 to 2008-09. It is observed that east as well as north-east zones received a little percentage share of subsidies during pre as well as post-liberalisation periods. East zone has got forth rank and north-east zone has got fifth during pre as well as post-liberalisation periods.

In south zone, subsidies of fertilizers have increased from Rs.132.06 crores in 1980-81 to Rs.26,229.29 crores in 2008-09. In west zone, these have increased from Rs.97.59 crores in 1980-81 to Rs.3,311.62 crores in 2000-01 and further increased to Rs.29,371.50 crores in 2008-09. In north zone, subsidies of fertilizers have increased from Rs.190.98 crores in 1980-81 to Rs.1,534.05 crores in 1990-91 and further increased to Rs.30,081.32 crores in 2008-09. On the hand in east zone, these have risen up from Rs.49.39 crores in 1980-81 to Rs.1,965.00 crores in 2000-01 and further increased to Rs.14,454.20 crores in 2008-09 and these have increased from Rs.1.87 crores in 1980-81 to Rs.147.29 crores in 2000-01 and further increased to Rs.1,044.36 crores in 2008-09 in north-east zone.

The percentage share of south zone has increased from 27.99 in 1980-81 to 28.28 in 1990-91 and declined to 25.92 in 2008-09. The percentage share of west zone has increased from 20.68 in 1980-81 to 22.77 in 1990-91 and declined to 24.09 in 1996-97 and again increased to 29.03 in 2008-09, whereas in north zone, it has increased from 40.47 per cent in 1980-81 to 40.76 per cent in 1985-86 and declined to 32.44 per cent in 2000-01 and further declined to 29.73 per cent in 2008-09. On the other hand, the percentage share of east zone has increased from 10.47 in 1980-81 to 12.38 in 1990-91 and declined to 13.29 in 2005-06, whereas north-east zone has got 0.40 per cent, 0.49 per cent and 1.17 per cent in 1980-81, 1990-91 and 2005-06 respectively.

**Table 4.8: Zone-wise Distribution of Fertilizers Subsidies in India during 1980-81 to 2008-09 (In Rs. Crores)**

| *Years/Zones* | *1980-81* | *1985-86* | *1990-91* | *1996-97* | *2000-01* | *2005-06* | *2008-09* |
|---|---|---|---|---|---|---|---|
| South | 132.06<br>(27.99) | 473.82<br>(26.25) | 1,311.90<br>(28.28) | 2,666.73<br>(32.73) | 3,848.08<br>(28.04) | 4,763.93<br>(26.80) | 26,229.29<br>(25.92) |
| West | 97.59<br>(20.68) | 367.86<br>(20.38) | 1,195.49<br>(22.77) | 1,962.60<br>(24.09) | 3,311.62<br>(24.13) | 4,790.16<br>(26.95) | 29,371.50<br>(29.03) |
| North | 190.98<br>(40.47) | 735.70<br>(40.76) | 1534.05<br>(33.07) | 2,483.93<br>(30.48) | 4,452.07<br>(32.44) | 5,651.01<br>(31.79) | 30,081.32<br>(29.73) |
| East | 49.39<br>(10.47) | 220.81<br>(12.23) | 574.19<br>(12.38) | 991.94<br>(12.17) | 1,965.00<br>(14.32) | 2,361.59<br>(13.29) | 14,454.20<br>(14.29) |
| North-East | 1.87<br>(0.40) | 6.60<br>(0.37) | 22.93<br>(0.49) | 43.22<br>(0.53) | 147.29<br>(1.07) | 207.70<br>(1.17) | 1,044.36<br>(1.03) |
| India | 471.88<br>(100) | 1,804.8<br>(100) | 4.638.56<br>(100) | 8,148.41<br>(100) | 13,724.05<br>(100) | 17,774.38<br>(100) | 10,1180.68<br>(100) |

*Source*: (1) Government of India, Fertilizers Association, Fertilizer Statistics, various issues, New Delhi.

(2) Government of India, Economic survey and union budget, various years.

*Note*: (1) Fertilizers subsidies of zones are calculated by adding the fertilizers subsidies on zone basis.

(2) Percentages are shown in parentheses.

At national level as well as zone level, fertilizers subsidies have increased in absolute terms and a lot of variation is seen in percentage share during pre as well as post-liberalisation periods. The fertilizers subsidies have increased the maximum times *i.e.* 45.55 times in north-east zone, whereas increased minimum times *i.e.* near about twenty in north zone in 2008-09 as compared to 1990-91. As compared to east zone, during pre-liberalisation period (1990-91), the south zone has got 2.28 times more of fertilizers subsidies, whereas in 2008-09, it has received 1.81 times more of fertilizers subsidies.

The subsidies of fertilizers in different states in south zone in India during 1980-81 to 2008-09 are shown in Table 4.9. This table reveals that in Andhra Pradesh these have increased from Rs.50.34 crores in 1980-81 to Rs.1,786.79 crores in 2000-01 and further increased to Rs.12,473.73 crores in 2008-09. In Karnataka, subsidies of fertilizers have risen up by 287.53 per cent in 1985-86,165.25 per cent in 1990-91, 36.86 per cent in 1996-97, 161.38 per cent in 2000-01, 20.93 per cent in 2005-06 and 455.37 per cent in 2008-09.

Kerala has received Rs.8.54 crores in 1980-81 as subsidies of fertilizers, whereas these have increased to Rs.29.97 in pre-liberalisation period (1985-86) and further increased to Rs.1,059.79 in post-liberalisation period (2008-09). Tamil Nadu has got

Rs.43.04 crores in 1980-81, which has increased by 226.29 per cent in 1985-86,120.00 per cent, 31.39 per cent, 94.91 per cent, 22.05 per cent and 432.15 per cent in 1990-91, 1996-97, 2000-01, 2005-06 and 2008-09 as compared to 1980-81, 1985-86, 1990-91, 1996-97, 2000-01 and 2005-06 respectively.

In Pondicherry, these subsidies of fertilizers have increased from Rs.11.41 crores in 1996-97 to Rs.113.49 crores in 2008-09, whereas these have increased from Rs.0.22 crores in 1996-97 to Rs.0.52 crores in 2005-06 and further increased to Rs.2.44 crores in 2008-09 in Andaman and Nicobar Islands.

The percentage share of Andhra Pradesh has increased from 10.67 in 1980-81 to 12.98 in 1990-91 and declined to 11.14 in 1996-97 and again increased to 12.33 in 2008-09. Karnataka has got third rank by receiving 6.39 per cent of fertilizers subsidies at the country level in 1980-81. It is found that in the same state, the percentage share has increased from 6.47 in 1985-86 to 8.07 in 2000-01 and declined to 7.54 in 2005-06. Kerala has got a little amount of percentage share during pre as well as post-liberalisation periods.

**Table 4.9: State-wise Distribution of Fertilizers Subsidies in South Zone in India during 1980-81 to 2008-09 (In Rs. Crores)**

| *Years/States* | *South Zone* | | | | | | |
|---|---|---|---|---|---|---|---|
| | *1980-81* | *1985-86* | *1990-91* | *1996-97* | *2000-01* | *2005-06* | *2008-09* |
| Andhra Pradesh | 50.34<br>(10.67) | 186.64<br>(10.34) | 602.31<br>(12.98) | 907.81<br>(11.14) | 1,786.79<br>(13.02) | 2,242.61<br>(12.62) | 12,473.73<br>(12.33) |
| Karnataka | 30.13<br>(6.39) | 116.76<br>(6.47) | 309.72<br>(6.68) | 423.87<br>(5.20) | 1,107.91<br>(8.07) | 1,339.76<br>(7.54) | 7,440.62<br>(7.35) |
| Kerala | 8.54<br>(1.81) | 29.97<br>(1.66) | 90.87<br>(1.96) | 96.27<br>(1.18) | 142.32<br>(1.04) | 177.88<br>(1.00) | 1,059.79<br>(1.05) |
| Tamil Nadu | 43.04<br>(9.12) | 140.45<br>(7.78) | 308.99<br>(6.66) | 405.97<br>(4.98) | 791.27<br>(5.77) | 965.75<br>(5.43) | 5,139.22<br>(5.08) |
| Pondicherry | – | – | – | 11.41<br>(0.14) | 19.45<br>(0.14) | 37.41<br>(0.21) | 113.49<br>(0.11) |
| Andaman and Nicobar Islands | – | – | – | 0.22<br>(0.003) | 0.35<br>(0.003) | 0.52<br>(0.003) | 2.44<br>(0.002) |
| Lakshadweep | – | – | – | 821.18<br>(10.08) | – | – | – |

*Source*: (1) Government of India, Fertilizers Association, Fertilizer Statistics, various issues, New Delhi.

(2) Government of India, Economic survey, union Budget, various years.

*Note*: (1) Fertilizers subsidies are calculated by multiplying the consumption of fertilizers (in 000 tonnes) with subsidy per tonne at national level.

(2) Percentages are shown in parentheses.

It is observed that the percentage share of Kerala has increased from 1.81 in 1980-81 to 1.96 in 1990-91 and declined to 1.05 in 2008-09. Tamil Nadu ranked second during 1980-81 to 1985-86, whereas it has lost its rank during 1990-91 to 2008-09. In Tamil Nadu, the percentage share has declined from 9.12 in 1980-81 to 4.98 in

1996-97 and increased to 5.77 in 2000-01. Pondicherry has got 0.14 per cent during 1996-97 to 2000-01 and 0.11 per cent in 2008-09, whereas Andaman and Nicobar Islands has got 0.003 per cent during 1996-97 to 2005-06 and 0.002 per cent in 2008-09. In 1996-97, Lakshadweep received Rs.821.18 crores (10.08 per cent at national level) as fertilizers subsidies.

The above table indicates that in all the states of south zone, the fertilizers subsidies have increased in absolute terms during the study period. As the year 2008-09 is compared to the year 1990-91, it is found that in Karnataka, the fertilizers subsidies have increased by twenty four times more of fertilizers subsidies, in Andhra Pradesh near about twenty one times and in Kerala near about twelve times. Andhra Pradesh has got 1.94 times fertilizers subsidies as compared to Karnataka during pre as well as post-liberalisation periods. Tamil Nadu has received 3.4 times more of fertilizers subsidies and 4.85 times in 1990-91 and 2008-09 respectively as compared to Kerala.

**Table 4.10: State-wise Distribution of Fertilizers Subsidies in West Zone in India during 1980-81 to 2008-09 (In Rs. Crores)**

| *Years/States* | *West Zone* | | | | | | |
|---|---|---|---|---|---|---|---|
| | *1980-81* | *1985-86* | *1990-91* | *1996-97* | *2000-01* | *2005-06* | *2008-09* |
| Gujarat | 31.27<br>(6.63) | 88.54<br>(4.91) | 262.67<br>(5.66) | 417.54<br>(5.12) | 616.78<br>(4.49) | 1,124.51<br>(6.33) | 6,974.23<br>(6.89) |
| Madhya Pradesh | 17.24<br>(3.65) | 90.77<br>(5.03) | 302.08<br>(7.77) | 499.44<br>(6.13) | 789.43<br>(5.75) | 826.61<br>(4.65) | 5,781.73<br>(5.71) |
| Chhattisgarh | – | – | – | – | – | 328.87<br>(1.85) | 1,879.94<br>(1.86) |
| Maharashtra | 36.89<br>(7.82) | 140.62<br>(7.79) | 489.87<br>(10.56) | 682.27<br>(8.37) | 1,353.38<br>(9.86) | 1,729.37<br>(9.73) | 10,423.32<br>(10.30) |
| Rajasthan | 11.84<br>(2.51) | 46.42<br>(2.57) | 137.97<br>(2.97) | 359.51<br>(4.41) | 546.26<br>(3.98) | 774.96<br>(4.36) | 4,273.22<br>(4.22) |
| Goa | 0.36<br>(0.08) | 1.51<br>(0.08) | 2.90<br>(0.06) | 3.08<br>(0.04) | 4.80<br>(0.03) | 4.85<br>(0.03) | 33.02<br>(0.03) |
| Daman and Diu | – | – | – | 0.18<br>(0.002) | 0.21<br>(0.002) | – | 1.58<br>(0.002) |
| Dadra Nagar Haveli | – | – | – | 0.57<br>(0.01) | 0.76<br>(0.01) | 0.99<br>(0.01) | 4.47<br>(0.004) |

*Source*: (1) Government of India, Fertilizers Association, Fertilizer Statistics, various issues, New Delhi.

(2) Government of India, Economic survey, union Budget, various years.

*Note*: (1) Fertilizers subsidies are calculated by multiplying the consumption of fertilizers (in 000 tonnes) with subsidy per tonne at national level.

(2) Percentages are shown in parentheses.

The fertilizers subsidies of different states of west zone in India during 1980-84 to 2008-09 are shown in Table 4.10. This table indicates that Gujarat State has got Rs.31.27 crores of fertilizers subsidies in 1980-81, which have increased to Rs.262.67 crores in 1990-91 and further increased to Rs.6,974.23 crores in 2008-09. In Madhya

Pradesh, these have increased from Rs.17.24 crores in 1980-81 to Rs.789.43 crores in 2000-01 and further increased to Rs.5,781.73 crores in 2008-09. Chhattisgarh has got Rs.328.87 crores and Rs.1,879.94 crores in 2005-06 and 2008-09 respectively.

In Maharashtra, these have increased from 36.89 crores in 1980-81 to Rs.10,423.32 crores in 2008-09. In Rajasthan these have risen up at decreasing rate by 291.21 percent, 197.19 percent, 160.58 percent, 51.94 percent, 41.87 percent 1990-91, 1996-97, 2000-01 except in 2008-09, it is 451.41 percent as compared to 2005-06.

Goa has received Rs. 0.36 crores in 1980-81, increased by 321.24 per cent in 1985.86 (from 1980-81) and further increased by 580.93 per cent in 2008-09 (from 2005-06). Dadra Nagar Haveli has received Rs.0.51 crores, Rs.0.76 crores, Rs.0.99 crores and Rs.4.47 crores in 1996-97, 2000-01, 2005-06 and 2008-09 respectively.

As percentage-wise analysis reveals that Maharashtra occupied topmost position among all the other states of the same zone throughout the study period. In Maharashtra, the percentage share has increased from 7.82 in 1980-81 to 12.60 in 1990-91 and declined to 10.30 in 2008-09. In Gujarat, it has declined from 6.63 per cent in 1980-81 to 4.91 per cent in 1985-86 and increased to 6.89 per cent in 2008-09, whereas the percentage share of subsidies have increased from 3.65 in 1980-81 to 6.51 in 1990-91 and declined to 5.71 in 2008-09 in Madhya Pradesh.

Chhattisgarh has received 1.85 per cent and 1.86 per cent in 2005-06 and 2008-09 respectively. Rajasthan has enjoyed a little percentage share (2.51 per cent) in 1980-81, increased to 4.41 in 1996-97 and declined to 3.98 in 2000-01 and again increased to 4.36 in 2005-06. During study, it is found that Goa has got 0.08 per cent during 1980-81 to 1985-86 and 0.03 per cent during 2000-01 to 2008-09, whereas Dadra Nagar Haveli received 0.01 per cent and 0.004 per cent during 1996-97 to 2005-06 and 2008-09 respectively.

From the above table, it is concluded that in all the states of west zone, the fertilizers subsidies in absolute terms has increased during post-liberalisation period as compared to pre-liberalisation period. Whereas a lot of variation is seen in percentage share during the study period. As the year 2008-09 is compared to the year 1990-91 of pre-liberalisation period, it is found that in Rajasthan, the fertilizers subsidies have increased more than thirty times, in Gujarat twenty seven times, in Madhya Pradesh near about nineteen times and in Maharashtra these have risen up twenty one times. Maharashtra has got 1.86 times more of fertilizers subsidies in 1990-91 and 1.5 times in 2008-09 as compared to Gujarat.

The fertilizers subsidies of different states of north zone of India during 1980-81 to 2008-09 are shown in Table 4.11. In Haryana, the fertilizers subsidies have increased from Rs.20.18 crores in 1980-81 to Rs.764.40 crores in 2000-01 and further increased to Rs.5,235.81 crores in 2008-09. In Punjab, these have gone up from Rs.66.76 crores in 1980-81 to Rs.454.03 crores in 1990-91 and further increased to Rs.7,181.47 crores in 2008-09. The fertilizers subsidies have risen up from Rs.100.81 crores in 1980-81 to Rs.2,521.17 crores in 2000-01 and further increased to Rs.16,380.73 crores in 2008-09 in Uttar Pradesh.

Jammu & Kashmir has got Rs.1.81 crores as fertilizers subsidies in 1980-81, gone up to Rs.15.84 crores in 1990-91 and further gone up to Rs.426.91 crores in 2008-09.

In Delhi, the subsidies have declined during pre and post-liberalisation periods except in 2008-09. Whereas in Himachal Pradesh these have increased from Rs. 1.42 crores in 1980-81 to Rs.12.87 crores in 1990-91 and further increased to Rs.233.03 in 2008-09. Uttarakhand has got Rs.10.60 crores and Rs.620.78 crores in 2005-06 and in 2008-09 respectively.

**Table 4.11: State-wise Distribution of Fertilizers Subsidies in North Zone in India during 1980-81 to 2008-09 (In Rs. Crores)**

| *Years/States* | *North Zone* | | | | | | |
|---|---|---|---|---|---|---|---|
| | *1980-81* | *1985-86* | *1990-91* | *1996-97* | *2000-01* | *2005-06* | *2008-09* |
| Haryana | 20.18 (4.28) | 78.18 (4.33) | 218.02 (4.70) | 390.81 (4.80) | 764.40 (5.57) | 991.91 (5.58) | 5,235.81 (5.17) |
| Punjab | 66.76 (14.15) | 230.75 (12.79) | 454.03 (9.79) | 619.99 (7.61) | 1,079.68 (7.87) | 1,481.28 (8.33) | 7,181.47 (7.10) |
| Uttar Pradesh | 100.81 (21.36) | 414.41 (22.96) | 833.29 (17.96) | 1,421.04 (17.44) | 2,521.17 (18.37) | 3,043.62 (17.12) | 16,380.73 (16.19) |
| Jammu & Kashmir | 1.81 (0.38) | 7.40 (0.41) | 15.84 (0.34) | 22.24 (0.27) | 53.39 (0.39) | 81.06 (0.46) | 426.91 (0.42) |
| Delhi | – | – | – | 12.17 (0.15) | 4.21 (0.03) | 0.40 (0.002) | 2.60 (0.003) |
| Himachal Pradesh | 1.42 (0.30) | 4.96 (0.27) | 12.87 (0.28) | 17.68 (0.22) | 29.21 (0.21) | 42.15 (0.24) | 233.03 (0.23) |
| Uttarakhand | – | – | – | – | – | 10.60 (0.06) | 620.78 (0.61) |

*Source*: (1) Government of India, Fertilizers Association, Fertilizer Statistics, various issues, New Delhi.

(2) Government of India, Economic survey, union Budget, various years.

*Note*: (1) Fertilizers subsidies are calculated by multiplying the consumption of fertilizers (in 000 tonnes) with subsidy per tonne at national level.

(2) Percentages are shown in parentheses.

It is observed that Uttar Pradesh is ahead by receiving maximum percentage share of fertilizers subsidies at national level among all the other six states of north zone during the pre and post-liberalisation periods. Punjab has got second position during pre as well as post-liberalisation periods followed by Haryana, Jammu & Kashmir, Himachal Pradesh, Uttarakhand and Delhi. Haryana's percentage share has increased from 4.28 in 1980-81 to 4.70 in 1990-91 and declined to 5.17 in 2008-09. In Punjab, it has declined from 14.15 in 1980-81 to 12.79 in 1985-86 and further declined to 7.61 per cent in 1996-97 and increased to 7.87 per cent in 2000-01, it has increased to 8.33 per cent in 2005-06. It has increased from 21.36 per cent in 1980-81 to 22.96 per cent in 1985-86 and declined to 17.96 per cent in 1990-91 and again increased to 18.37 per cent in 2000-01 in Uttar Pradesh.

In Himachal Pradesh, its percentage share has increased from 0.28 in 1980-81 to 0.33 in 1990-91 and declined to 0.23 in 2008-09, whereas Jammu & Kashmir has got 0.38 per cent, 0.41 per cent, 0.39 per cent and 0.42 per cent in 1980-81, 1996-97,

2000-01 and 2008-09 respectively. It is found that in Delhi, the percentage share at country level has declined during pre as well as post-liberalisation periods.

The fertilizers subsidies have increased in absolute terms in all the states of north zone (except in Delhi) during the pre as well post-liberalisation periods, whereas a lot of variation is seen in percentage-wise analysis. As the year 2008-09 is compared to the year 1990-91, it is found that in Uttar Pradesh these have increased by 19.66 times, Jammu & Kashmir near about twenty seven times, Haryana twenty four times, Himachal Pradesh eighteen times and Punjab fifteen times. In 1990-91, Punjab has got two times more of fertilizers subsidies and 1.37 times in 2008-09 than that of Haryana, whereas Uttar Pradesh has received 52.61 times more of fertilizers subsidies in 1990-91, whereas 38.37 times more of fertilizers subsidies in 2008-09 than that of Jammu & Kashmir.

The subsidies of fertilizers of east zone of India during 1980-81 to 2008-09 is shown in Table 4.12. This table shows that in Bihar, these have increased from Rs.17.92 in 1980-81 to Rs. 222.54 in 1990-91 and further increased to Rs.5,512.02 crores in 2008-09. Jharkhand has got Rs.118.12 crores and Rs.598.20 crores in 2005-06 and 2008-09 respectively. In West Bengal, these have risen up from Rs.24.78 crores in 1980-81 and further increased to Rs.6,171.27 crores in 2008-09, whereas in Odisha, these subsidies have gone up by 341.12 per cent in 1985-86,142.64 per cent in 1990-91,79.64 per cent in 1996-97,103.79 per cent in 2000-01,32.27 per cent in 2005-06 and 526.27 per cent in 2008-09 as compared to the predecessor time in given table.

**Table 4.12: State-wise Distribution of Fertilizers Subsidies in East Zone in India during 1980-81 to 2008-09 (In Rs. Crores)**

| *Years/States* | *East Zone* | | | | | | |
|---|---|---|---|---|---|---|---|
| | *1980-81* | *1985-86* | *1990-91* | *1996-97* | *2000-01* | *2005-06* | *2008-09* |
| Bihar | 17.92<br>(3.80) | 105.39<br>(5.84) | 222.54<br>(4.80) | 403.32<br>(4.95) | 811.12<br>(5.91) | 807.38<br>(4.54) | 5,512.02<br>(5.45) |
| Jharkhand | – | – | – | – | – | 118.12<br>(0.66) | 598.20<br>(0.59) |
| Odisha | 6.69<br>(1.42) | 29.53<br>(1.64) | 71.65<br>(1.54) | 128.70<br>(1.58) | 262.29<br>(1.91) | 346.93<br>(1.95) | 2,172.72<br>(2.15) |
| West Bengal | 24.78<br>(5.25) | 85.89<br>(4.76) | 280.01<br>(6.04) | 459.91<br>(5.64) | 891.59<br>(6.50) | 1,089.15<br>(6.13) | 6,171.27<br>(6.10) |

*Source*: (1) Government of India, Fertilizers Association, Fertilizer Statistics, various issues, New Delhi.

(2) Government of India, Economic survey, union Budget, various years.

*Note*: (1) Fertilizers subsidies are calculated by multiplying the consumption of fertilizers (in 000 tonnes) with subsidy per tonne at national level.

(2) Percentages are shown in parentheses.

As percentage-wise analysis shows that Bihar has got 3.80 per cent of fertilizers subsidies at country level in 1980-81, whereas Odisha has received 1.42 per cent and West Bengal has got 5.25 per cent in the same year. In 1990-91, Bihar, Odisha and West Bengal have received 4.80 per cent, 1.54 per cent and 6.04 per cent respectively.

Bihar's percentage share has increased from 4.95 in 1996-97 to 5.91 in 2000-01 and declined to 4.54 in 2005-06 and again increased to 5.45 in 2008-09. In Jharkhand, the percentage share has declined from 0.66 in 2005-06 to 0.59 in 2008-09. In Odisha, the percentage share has increased from 1.58 in 1996-97 to 2.15 in 2008-09. It is observed that West Bengal is ahead among all the other states like Bihar, Jharkhand and Odisha during 1990-91 to 2008-09 by consuming maximum amount of fertilizers subsidies. It has got 5.64 per cent, 6.13 per cent and 6.10 per cent in 2000-01, 2005-06 and 2008-09 respectively.

In all the states of east zone, the fertilizers subsides have been increased in absolute terms during the study period. As post-liberalisation period (2008-09) is compared to pre-liberalisation period (1990-91), it is found that in Odisha, these have increased more than thirty times, in Bihar twenty four times and in West Bengal twenty two times. As state-wise analysis shows that West Bengal has got 1.26 times more of fertilizers subsidies and 1.11 times in 1990-91 and 2008-09 respectively as compared to Bihar.

**Table 4.13: State-wise Distribution of Fertilizers Subsidies in North-East Zone in India during 1980-81 to 2008-09 (In Rs. Crores)**

| *Years/States* | *North-East Zone* | | | | | | |
|---|---|---|---|---|---|---|---|
| | *1980-81* | *1985-86* | *1990-91* | *1996-97* | *2000-01* | *2005-06* | *2008-09* |
| Assam | 0.81<br>(0.17) | 3.51<br>(0.19) | 14.01<br>(0.30) | 28.64<br>(0.35) | 115.54<br>(0.84) | 174.06<br>(0.98) | 896.10<br>(0.89) |
| Tripura | 0.17<br>(0.04) | 1.08<br>(0.06) | 3.43<br>(0.07) | 4.42<br>(0.05) | 7.56<br>(0.06) | 13.33<br>(0.07) | 56.34<br>(0.06) |
| Manipur | 0.26<br>(0.06) | 1.02<br>(0.06) | 3.06<br>(0.07) | 7.09<br>(0.09) | 18.11<br>(0.13) | 12.48<br>(0.07) | 52.28<br>(0.05) |
| Meghalaya | 0.23<br>(0.05) | 0.63<br>(0.03) | 0.99<br>(0.02) | 1.76<br>(0.02) | 3.17<br>(0.02) | 4.31<br>(0.02) | 14.99<br>(0.01) |
| Nagaland | 0.01<br>(0.001) | 0.05<br>(0.003) | 0.41<br>(0.01) | 0.44<br>(0.01) | 0.33<br>(0.002) | 0.54<br>(0.003) | 3.70<br>(0.004) |
| Mizoram | 0.01<br>(0.001) | 0.02<br>(0.001) | 0.31<br>(0.01) | 0.20<br>(0.002) | 1.19<br>(0.01) | 2.00<br>(0.01) | 17.67<br>(0.02) |
| Sikkim | 0.37<br>(0.08) | 0.25<br>(0.01) | 0.57<br>(0.01) | 0.38<br>(0.005) | 0.89<br>(0.01) | 0.31<br>(0.002) | – |
| Arunachal Pradesh | 0.01<br>(0.001) | 0.04<br>(0.002) | 0.16<br>(0.004) | 0.28<br>(0.003) | 0.50<br>(0.004) | 0.66<br>(0.004) | 3.29<br>(0.003) |

*Source*: (1) Government of India, Fertilizers Association, Fertilizer Statistics, various issues, New Delhi.

(2) Government of India, Economic survey, union Budget, various years.

*Note*: (1) Fertilizers subsidies are calculated by multiplying the consumption of fertilizers (in 000 tonnes) with subsidy per tonne at national level.

(2) Percentages are shown in parentheses.

The fertilizers subsidies in north-east zone of India during 1980-81 to 2008-09 is shown in Table 4.13. In Assam, the fertilizers subsidies have increased from Rs.0.81 crores in 1980-81 to Rs.14.01 crores in 1990-91 and further increased to Rs.896.10

crores in 2008-09. In Tripura, these have increased from Rs.0.17 crores in 1980-81 to Rs.4.42 crores in 1996- 97 and further increased to Rs.56.34 crores in 2008-09. In Manipur, these have gone up by 286.99 per cent in 1985-86, 131.88 per cent in 1996-97, 155.49 per cent in 2000-01 and declined by 31.11 per cent in 2005-06 and again increased by 319.03 per cent in 2008-09 as compared to 1980-81, 1985-86, 1990-91, 1996-97 and 2000-01 respectively.

In Nagaland, fertilizers subsidies have increased from Rs.0.01 crores in 1980-81 to Rs.0.41 crores in 1985-86 and declined to Rs.0.33 crores in 2000-01 and again increased to Rs.3.70 crores in 2008-09, whereas in Mizoram, these have gone up from Rs.0.01 crores in 1980-81 to Rs.0.31 crores in 1990-91 and further gone up to Rs.17.67 crores in 2008-09. It is found that in Sikkim, the fertilizers subsidies have declined in all the study years except in 1990-91 and 2000-01. On the other hand, these have gone up from Rs.0.01 crores in 1980-81 to Rs.0.16 crores in 1990-91 and further increased to Rs.3.29 crores in 2008-09 in Andhra Pradesh.

The percentage share of Assam has increased from 0.17 in 1980-81 to 0.89 in 2008-09 (more than five times). The percentage share of Tripura has increased from 0.04 in 1980-81 to 0.06 in 1985-86 and declined to 0.05 in 1996-97 and again increased to 0.07 in 2005-06, whereas Mizoram has got 0.001 per cent, 0.01 per cent and 0.02 per cent in 1980-81 to 1985-86, 2000-01 to 2005-06 and 2008-09 respectively. Manipur received 0.06 percentage share at national level in 1980-81, increased to 0.09 and declined to 0.05 in 2008-09. In Meghalaya, the percentage share has declined from 0.05 in 1980-81 to 0.01 in 2008-09, whereas in Nagaland it has increased from 0.001 per cent in 1980-81 to 0.003 per cent in 1985-86 and further increased to 0.004 per cent in 2008-09. During 1980-81 to 2008-09, it is observed that Sikkim and Arunachal Pradesh enjoyed a little percentage share of fertilizers subsidies.

It is observed that in Assam, Tripura, Meghalaya and Arunachal Pradesh states the fertilizers subsidies have increased in absolute terms, whereas a lot of variation is found in percentage-wise percentage share among all the states of north-east zone during pre and post-liberalisation period. While the year 2008-09 is compared to the year 1990-91, in Assam, these have increased by 63.96 times, in Mizoram fifty seven, in Arunachal Pradesh more than twenty, in Manipur seventy, in Tripura more than sixteen and in Nagaland only nine times. In 1990-91, Assam has got four times more of fertilizers subsidies as compared to Tripura, whereas in 2008-09, it has received approximately sixteen times. Nagaland has got 2.56 times more of fertilizers subsidies and 1.12 times in 1990-91 and 2008-09 respectively as compared to Arunachal Pradesh.

As different zones as well as states have different levels of gross cropped area, it may be more meaningful to analyse the fertilizers subsidies per hectare of GCA among all the zones and states. The fertilizers subsidies per hectare of different zones in India, during 1980-81 to 2005-06 are shown in Table 4.14. This table reveals that in India the per hectare fertilizers subsidies have increased from Rs.901.82 in 1980-81 to Rs.3,241.64 in 1985-86 and further increased to Rs.22,309.86 in 1996-97. It is observed that at country level, these have risen by 126.18 per cent in 2005-06 as compared to 2000-01. In south zone, these have increased from Rs.166.37 in 1980-81 to Rs.569.11

in 1985-86 and further increased to Rs.20,974.65 in 2005-06. In west zone has got Rs.88.18, Rs.863.11 and Rs.8,119.40 in 1980-81, 1990-91 and 2005-06 respectively.

North zone has got Rs.210.14, Rs.1,568, Rs.4,275.31 and Rs.10,233.87 in 1980-81, 1990-91, 1996-97 and 2005-06 respectively. In east zone, these have increased by 323.91 per cent in 1985-86, 73.82 per cent in 1996-97, 100.30 per cent in 2000-01 and 190.88 per cent in 2005-06 as compared to 1980-81, 1985-96, 1996-97 and 2000-01, on the other hand these have increased from Rs.34.80 in 1980-81 to Rs.427.63 in 1990-91 and further increased to Rs.2,589.75 in 2005-06 in north-east zone.

**Table 4.14: Zone-wise distribution of Fertilizers Subsidies in India during 1980-81 to 2008-09 (In Rs./Hectare)**

| *Years/Zones* | *1980-81* | *1985-86* | *1990-91* | *1996-97* | *2000-01* | *2006-07* |
|---|---|---|---|---|---|---|
| South | 166.37<br>(29.92) | 569.11<br>(28.52) | 1,479.36<br>(29.91) | 5,591.12<br>(42.06) | 9,890.68<br>(45.97) | 20,974.65<br>(43.64) |
| West | 88.18<br>(10.58) | 290.80<br>(14.57) | 863.11<br>(17.45) | 1,710.18<br>(12.87) | 2,804.06<br>(13.03) | 8,119.40<br>(16.89) |
| North | 210.14<br>(37.80) | 749.42<br>(37.55) | 1,568.00<br>(31.71) | 4,275.31<br>(32.16) | 5,056.84<br>(23.50) | 10,233.87<br>(21.29) |
| East | 56.49<br>(10.16) | 239.46<br>(12.00) | 607.14<br>(12.28) | 1,055.35<br>(7.94) | 2,113.81<br>(9.82) | 6,148.76<br>(12.79) |
| North-East | 34.80<br>(6.26) | 146.74<br>(7.35) | 427.63<br>(8.65) | 660.62<br>(4.97) | 1,652.16<br>(7.68) | 2,589.75<br>(5.39) |
| India | 555.98<br>(100) | 1,995.53<br>(100) | 4,945.25<br>(100) | 13,292.58<br>(100) | 21,517.56<br>(100) | 48,066.43<br>(100) |

*Source*: (1) Government of India Fertilizers Association, Fertilizer Statistics, various issues, New Delhi.

(2) Government of India, Economic survey, union Budget, various years.

(3) Government of Punjab, Statistical Abstract, Various years.

*Note*: (1) Fertilizers subsidies per hectare of zones are calculated by adding the fertilizers subsidies per hectare on zone basis.

(2) Percentages are shown in parentheses.

South zone received 29.92 per cent of fertilizers subsidies per hectare in 1980-81, increased to 29.91 per cent in 1990-91 and further increased to 42.06 per cent in 1996-97 and declined to 43.64 per cent in 2005-06, whereas the percentage share of fertilizers subsidies have increased from 10.58 in 1980-81 to 17.45 in 1990-91 and declined to 12.87 in 1996-97 and again increased to 13.03 in 2000-01 and further increased to 16.89 in 2005-06 in west zone.

On the other hand, in north zone, the percentage share has declined from 37.80 in 1980-81 to 31.71 in 1990-91 and further declined to 21.29 in 2006-07. In east zone, this percentage share has increased from 10.16 in 1980-81 to 12.00 in 1985-86 and declined to 7.94 in 1990-91 and again increased to 12.79 in 2005-06, whereas north-east has got 6.26 per cent, 7.35 per cent, 8.65 per cent, 4.97 per cent, 7.68 per cent and 5.39 per cent in 1980-81, 1985-86, 1990-91, 1996-97, 2000-01 and 2005-06 respectively.

In India as well as in the entire five zones, the fertilizers subsidises Rs. per hectare have increased in absolutes terms during pre as well as post-liberalisation periods. As post-liberalisation period (2006-07) is compared to pre liberalisation period (1990-91), it is observed that in south zone, the fertilizers subsidies have increased by fourteen times, in east zone ten times, in west zone more than nine times, in north zone more than six times and in north-east zone six times. As compared to west zone, south zone has got 1.71 times more of fertilizers subsidies in 1990-91 and 2.58 times in 2006-07, whereas north zone has got near about three times more of fertilizers subsidies in 1990-91 and 1.66 times in 2006-07 as compared to east zone.

The fertilizers subsidies per hectare in different states south zone of India during 1980-81 to 2006-07 is shown in Table 4.15. During 1980-81 to 1985-86, Tamil Nadu has got first position, whereas in remaining years it has lost its same position. Tamil Nadu has got Rs.66.83, Rs.463.58, Rs.1,245.92 and 1,615.37 in 1980-81, 1990-91, 2000-01 and 2006-07 respectively. In Andhra Pradesh, these have gone up by 274.35 per cent in 1985-86 194.74 per cent, 48.07 per cent, 95.07 per cent and 72.33 per cent in 1990-91 1996-97, 2000-01 and 2006-07 respectively. In Kerala, these have increased from Rs.29.98 in 1980-81 to Rs.299.41 in 1990-91 and further increased to Rs.837.68 in 2006-07.

**Table 4.15: State-wise distribution of Fertilizers Subsidies in South Zone in India during 1980-81 to 2006-07 (In Rs./Hectare)**

| *Years/States* | *South Zone* | | | | | |
|---|---|---|---|---|---|---|
| | *1980-81* | *1985-86* | *1990-91* | *1996-97* | *2000-01* | *2006-07* |
| Andhra Pradesh | 41.17<br>(7.41) | 154.13<br>(7.72) | 454.30<br>(9.19) | 672.69<br>(5.06) | 1,316.46<br>(6.12) | 2,268.67<br>(4.72) |
| Karnataka | 28.39<br>(5.11) | 104.68<br>(5.25) | 262.08<br>(5.30) | 341.47<br>(2.57) | 900.07<br>(4.18) | 1,397.64<br>(2.91) |
| Kerala | 29.98<br>(5.39) | 104.49<br>(5.24) | 299.41<br>(6.05) | 316.76<br>(2.38) | 469.98<br>(2.18) | 837.68<br>(1.74) |
| Tamil Nadu | 66.83<br>(12.02) | 205.81<br>(10.31) | 463.58<br>(9.37) | 624.76<br>(4.70) | 1,245.92<br>(5.79) | 1,615.37<br>(3.36) |
| Pondicherry | – | – | – | 3,334.50<br>(25.09) | 5,883.38<br>(27.34) | 14,345.50<br>(29.85) |
| Andaman and Nicobar Islands | – | – | – | 28.95<br>(0.22) | 74.87<br>(0.35) | 509.79<br>(1.06) |
| Lakshadweep | – | – | – | 272.00<br>(2.05) | – | – |

*Source*: (1) Government of India, Fertilizers Association, Fertilizer Statistics, various issues, New Delhi.

(2) Government of India, Economic survey, union Budget, various years.

(3) Government of Punjab, Statistical Abstract, Various years.

*Note*: (1) Fertilizers subsidies per hectare of states are calculated by multiplying the consumption of fertilizers (in kgs) per hectare with subsidy per kg.

(2) Percentages are shown in parentheses.

In Karnataka, subsidies of fertilizers have risen up from Rs.28.39 in 1980-81 to Rs.262.08 in 1980-81 to Rs.262.08 in 1990-91 and further increased to Rs.1,397.64 in 2006-07. In Pondicherry, these have increased from Rs.3,334.50 in 1996-97 to Rs.5,883.38 in 2000-01 and further increased to Rs.14,345.50 in 2006-07, whereas in Andaman and Nicobar Islands fertilizers subsidies have risen up by 158.65 per cent in 2000-01 and 580.90 per cent in 2006-07 as compared to 1996-97 and 2000-01 respectively.

This table reveals that Tamil Nadu has got first rank during 1980-81 to 1990-91, whereas Pondicherry occupied first position during 1996-97 to 2006-07. The percentage share of Andhra Pradesh has increased from 7.41 in 1980-81 to 7.72 in 1985-86 and declined to 5.06 in 1996-97 and further declined to 4.72 in 2006-07, whereas increased from 5.11 in 1980-81 to 5.25 in 1985-86 and declined to 2.57 in 1996-97 and again increased to 4.18 in 2000-01 in Karnataka.

Tamil Nadu has enjoyed top most rank in 1980-81 by getting 12.02 per cent at National level, whereas it has lost its position in 1996-97 by getting 4.70 per cent and has got third rank in 2000-01 by receiving 5.79 per cent in 2000-01 and again has got third position in 2006-07 by receiving 3.36 per cent in 2006-07. Pondicherry has got 25.09 per cent, 27.34 per cent and 29.85 per cent in 1996-97, 2000-01, and 2006-07 respectively. It is found that the percentage share of Andaman and Nicobar Islands has increased during pre as well as post-liberalisation periods.

From above table, it is found that in all the states of south zone the fertilizers subsidies in Rs. per hectare have increased in absolutes terms during 1980-81 to 2006-07, whereas the percentage share has increased during pre-liberalisation period in Andhra Pradesh, Karnataka and Kerala and a lot of variation is observed during post-liberalisation period. Among all the states of this zone, in Karnataka, these have risen up by more than five times, in Andhra Pradesh 4.99 times and in Kerala near about three times during post-liberalisation period (2006-07) from pre liberalisation period (1990-91). In 1990-91, Andhra Pradesh from Karnataka and Tamil Nadu from Kerala have received less than times more of fertilizers subsidies per hectare, the same pattern is also seen in post-liberalisation period.

The fertilizer subsidies per hectare of different states of west zone in India during 1980-81 to 2006-07 are shown in Table 4.16. Gujarat has got topmost rank during 1980-81 to 1996-97, as in the same state, these have increased from Rs.29.37 in 1980-81 to Rs.252.26 in 1990-91 and further increased to Rs.1,286.62 in 2006-07. Maharashtra has received Rs.18.28 in 1980-81, which have increased to Rs.223.03 in 1990-91 and further increased to Rs.3,243.73 in 2006-07. The fertilizers subsidies have increased from Rs.8.09 in 1980-81 to Rs.125.87 in 1990-91 and further increased to Rs.1,133.61 in 2006-07 in Madhya Pradesh.

In Rajasthan, these have increased from Rs.6.85 in 1980-81 to Rs.70.83 in 1990-91 and further increased to Rs.301.60 in 2006-07, whereas in Dadra Nagar Haveli, these have gone up by 32.50 per cent in 2000-01 as compared to 1996-97. In Goa, fertilizers subsidies have increased from Rs. 25.59 in 1980-81 to Rs.404.74 in 2006-07, whereas in 2000-01 in Daman and Diu these have risen up by 16.12 per cent.

It is observed that Gujarat has got topmost position during 1980-81 to 1996-97 followed by Maharashtra, Goa, Madhya Pradesh, Chhattisgarh, Daman and Diu and Dadra Nagar Haveli. The percentage share has declined from 5.28 in 1980-81 to 4.58 in 1985-86 and increased to 5.10 in 1990-91 and again declined to 2.68 in 2000-01 in Gujarat.

Madhya Pradesh's percentage share has increased from 1.46 in 1980-81 to 2.05 in 2000-01 and further increased to 2.36 in 2006-07. Maharashtra has got 3.29 per cent, 4.51 per cent, 2.86 per cent, 6.75 per cent in 1980-81, 1990-91, 2000-01 and 2006-07 respectively. The percentage share of Rajasthan at India level has increased from 1.23 in 1980-81 to 1.32 in 2000-01 and declined to 0.63 in 2006-07. In Goa, it has declined from 4.60 per cent in 1980-81 to 0.84 per cent in 2006-07, on the other hand it is found that the percentage share has declined from 1996-97 to 2000-01 in Daman and Diu and Dadra Nagar Haveli.

**Table 4.16: State-wise distribution of Fertilizers Subsidies in West Zone in India during 1980-81 to 2006-07 (In Rs./Hectare)**

| *Years/States* | *West Zone* | | | | | |
|---|---|---|---|---|---|---|
| | *1980-81* | *1985-86* | *1990-91* | *1996-97* | *2000-01* | *2006-07* |
| Gujarat | 29.37<br>(5.28) | 91.37<br>(4.58) | 252.26<br>(5.10) | 377.15<br>(2.84) | 575.79<br>(2.68) | 1,286.62<br>(2.68) |
| Madhya Pradesh | 8.09<br>(1.46) | 39.41<br>(1.97) | 125.87<br>(2.55) | 202.97<br>(1.53) | 440.86<br>(2.05) | 1,133.61<br>(2.36) |
| Chhattisgarh | – | – | – | – | – | 1,749.11<br>(3.64) |
| Maharashtra | 18.28<br>(3.29) | 68.42<br>(3.43) | 223.03<br>(4.51) | 312.11<br>(2.35) | 616.41<br>(2.86) | 3,243.73<br>(6.75) |
| Rajasthan | 6.85<br>(1.23) | 25.58<br>(1.28) | 70.83<br>(1.43) | 172.64<br>(1.30) | 283.49<br>(1.32) | 301.60<br>(0.63) |
| Goa | 25.59<br>(4.60) | 66.03<br>(3.31) | 191.13<br>(3.86) | 154.55<br>(1.16) | 280.19<br>(1.30) | 404.74<br>(0.84) |
| Daman and Diu | – | – | – | 262.29<br>(1.97) | 304.57<br>(1.42) | – |
| Dadra Nagar Haveli | – | – | – | 228.48<br>(1.72) | 302.74<br>(1.41) | – |

*Source*: (1) Government of India, Fertilizers Association, Fertilizer Statistics, various issues, New Delhi.

(2) Government of India, Economic survey, union Budget, various years.

(3) Government of Punjab, Statistical Abstract, Various years.

*Note*: (1) Fertilizers subsidies per hectare of states are calculated by multiplying the consumption of fertilizers (in kgs) per hectare with subsidy per kg.

(2) Percentages are shown in parentheses.

The fertilizers subsidies per hectare have been increased in the entire states of the west zone except in Goa throughout the study period, whereas percentage share analysis reveals a lot of variations. As the year 2006-07 is compared to the year 1990-

91, in Maharashtra, these have increased the maximum *i.e.* fourteen times, whereas in Goa, it has increased minimum *i.e.* only two times. During pre-liberalisation period, Gujarat has got near about four times more of fertilizers subsidies as compared to Madhya Pradesh and Maharashtra near about three times as compared to Rajasthan, whereas during post-liberalisation period (2006-07), Gujarat only 1.13 times more of fertilizers subsidies as compared to Madhya Pradesh and Maharashtra near about eleven times Rajasthan.

The fertilizers subsidies per hectare in north zone in India during 1980-81 to 2006-07 are shown in Table 4.17. Punjab is leading among all the other states during 1980-81 to 1990-91 and during 2000-01 to 2006-07 (Delhi is leading in 1996-97). In Punjab state these subsidies have risen up from Rs.98.06 in 1980-81 to Rs.590.76 in 1990-91 and further risen up to Rs.4,678.34 in 2006-07. In Haryana, the subsidies per hectare have increased from Rs.37.10 in 1980-81 to Rs.366.50 in 1990-91 and further increased to Rs.2,058.57 in 2006-07. Uttar Pradesh has got Rs.41.20, Rs.325.41, Rs.929.90 and Rs.1,689.73 in 1980-81, 1990-91, 2000-01 and 2005-06 respectively.

**Table 4.17: State-wise distribution of Fertilizers Subsidies in North Zone in India during 1980-81 to 2006-07 (In Rs./Hectare)**

| *Years/States* | *North Zone* | | | | | |
|---|---|---|---|---|---|---|
| | *1980-81* | *1985-86* | *1990-91* | *1996-97* | *2000-01* | *2006-07* |
| Haryana | 37.10<br>(6.67) | 139.48<br>(6.99) | 366.50<br>(7.41) | 639.36<br>(4.81) | 1,247.49<br>(5.80) | 2,058.57<br>(4.28) |
| Punjab | 98.06<br>(17.64) | 322.19<br>(16.15) | 590.76<br>(11.95) | 785.44<br>(5.91) | 1,356.48<br>(6.30) | 4,678.34<br>(9.73) |
| Uttar Pradesh | 41.20<br>(7.41) | 165.10<br>(8.27) | 325.41<br>(6.58) | 540.42<br>(4.07) | 929.90<br>(4.32) | 1,689.73<br>(3.52) |
| Jammu & Kashmir | 18.70<br>(3.36) | 71.77<br>(3.60) | 147.83<br>(2.99) | 205.18<br>(1.54) | 477.88<br>(2.22) | 893.92<br>(1.86) |
| Delhi | – | – | – | 1,919.38<br>(14.44) | 737.57<br>(3.43) | 282.98<br>(0.59) |
| Himachal Pradesh | 15.07<br>(2.71) | 50.88<br>(2.55) | 137.51<br>(2.78) | 185.53<br>(1.40) | 307.52<br>(1.43) | 605.14<br>(1.26) |
| Uttarakhand | – | – | – | – | – | 25.19<br>(0.05) |

*Source*: (1) Government of India, Fertilizers Association, Fertilizer Statistics, various issues, New Delhi.

(2) Government of India, Economic survey, union Budget, various years.

(3) Government of Punjab, Statistical Abstract, Various years.

*Note*: (1) Fertilizers subsidies per hectare of states are calculated by multiplying the consumption of fertilizers (in kgs) per hectare with subsidy per kg.

(2) Percentages are shown in parentheses.

In Jammu & Kashmir, these have gone up by 283.73 in 1985-86,105.98 in 1990-91,38.80 in 1996-97,132.90 in 2000-01 and 87.06 in 2006-07 as compared to 1980-81, 1985-86, 1990-91, 1996-97, 2000-01 and 2006-07 respectively. It is found that in Delhi, these have declined during 1996-97 to 2006-07. In Himachal Pradesh, subsidies per hectare have increased from Rs.15.07 in 1980-81 to Rs.137.51 in 1990-91 and further increased to Rs.605.14 in 2006-07, whereas Uttarakhand has got Rs.25.19 in post-liberalisation period (2006-07).

Percentage-wise analysis shows that Punjab has got first rank during the study period followed by Uttar Pradesh, Haryana, Jammu & Kashmir, Delhi, Himachal Pradesh and Uttarakhand. In Punjab, it has declined from 17.64 in 1980-81 to 5.91 per cent in 1996-97 and increased to 9.73 per cent in 2006-07. In Haryana it has increased from 6.67 per cent in 1980-81 to 6.99 per cent in 1985-86 and declined to 4.81 per cent in 1996-97 and again increased to 5.80 per cent in 2000-01, whereas Jammu & Kashmir has got 3.36 per cent, 2.99 per cent, 2.22 per cent and 1.86 per cent in 1980-81, 1990-91, 2000-01 and 2006-07 respectively.

From the above analysis, it is observed that in absolute terms, fertilizers subsidises have been increased, whereas variations are seen in entries states of north zone during the study period. As post-liberalisation period (2006-07) is compared to pre-liberalisation period (1990-91), in Punjab, these have risen up near about eight times, in Jammu & Kashmir six times, in Haryana near about six times, in Uttar Pradesh more than five times and in Himachal Pradesh more than four times. Punjab has got 1.61 times more of fertilizers subsidies and 2.27 times than that of Haryana in 1990-91 and in 2006-07 respectively, whereas during the same time reverse figures are observed in Uttar Pradesh when compared to Jammu & Kashmir.

**Table 4.18: State-wise Distribution of Fertilizers Subsidies in East Zone in India during 1980-81 to 2006-07 (In Rs./Hectare)**

| *Years/States* | *East Zone* | | | | | |
|---|---|---|---|---|---|---|
| | *1980-81* | *1985-86* | *1990-91* | *1996-97* | *2000-01* | *2006-07* |
| Bihar | 16.14<br>(2.90) | 100.14<br>(5.02) | 211.19<br>(4.27) | 395.20<br>(2.97) | 805.60<br>(3.74) | 1,654.23<br>(3.44) |
| Jharkhand | – | – | – | – | – | 613.15<br>(1.28) |
| Odisha | 7.69<br>(1.38) | 31.87<br>(1.60) | 74.30<br>(1.50) | 155.66<br>(1.17) | 332.26<br>(1.54) | 1,103.96<br>(2.30) |
| West Bengal | 32.66<br>(5.87) | 107.46<br>(5.38) | 321.65<br>(6.50) | 504.48<br>(3.80) | 975.95<br>(4.54) | 2,777.40<br>(5.78) |

*Source*: (1) Government of India, Fertilizers Association, Fertilizer Statistics, various issues, New Delhi.

(2) Government of India, Economic survey, union Budget, various years.

(3) Government of Punjab, Statistical Abstract, Various years.

*Note*: (1) Fertilizers subsidies per hectare of states are calculated by multiplying the consumption of fertilizers (in kgs) per hectare with subsidy per kg.

(2) Percentages are shown in parentheses.

The fertilizers subsidies per hectare in east zone of India during 1980-81 to 2006-07 are shown in Table 4.18. This table reveals that in all the states of this zone the fertilizers subsidies per hectare have been increased during pre and post-liberalisation periods at different rates. West Bengal is leading among all the other states followed by Bihar and Odisha during 1980-81 to 2006-07.

In West Bengal, these subsidies have increased from Rs.32.66 in 1980-81 to Rs.321.65 in 1990-91 and further increased to Rs.2,777.40 in 2006-07. In Bihar these have increased from Rs.16.14 in 1980-81 to Rs.211.19 in 1990-91 and further increased to Rs.1,654.23 in 2006-07. Whereas in Odisha, these have risen up by 314.54 per cent in 1985-86, 155.66 per cent in 1996-97, 113.45 per cent in 2000-01 and 232.26 per cent in 2006-07.

As percentage-wise analysis reveals that in Bihar, the percentage share has risen from 2.90 in 1980-81 to 5.02 in 1985-86 and declined to 4.27 in 1990-91 and further declined to 3.44 in 2006-07, whereas in Odisha, the percentage share has increased from 1.38 in 1980-81 to 1.60 in 1985-86 and declined to 1.50 in 1990-91 and again increased to 2.30 in 2006-07. Jharkhand has got 1.28 per cent in 2006-07, whereas West Bengal has received 5.87 per cent, 6.50 per cent, 4.54 per cent and 5.78 per cent in 1980-81, 1990-91, 2000-01 and 2006-07 respectively.

**Table 4.19: State-wise distribution of Fertilizers Subsidies in North-East Zone in India during 1980-81 to 2006-07 (In Rs./Hectare)**

| *Years/States* | *North-East Zone* | | | | | |
|---|---|---|---|---|---|---|
| | *1980-81* | *1985-86* | *1990-91* | *1996-97* | *2000-01* | *2006-07* |
| Assam | 2.37<br>(0.43) | 9.24<br>(0.46) | 36.71<br>(0.74) | 71.50<br>(0.54) | 283.66<br>(1.32) | 670.72<br>(1.40) |
| Tripura | 4.65<br>(0.84) | 25.57<br>(1.28) | 76.66<br>(1.55) | 95.99<br>(0.72) | 176.24<br>(0.82) | 624.25<br>(1.30) |
| Manipur | 12.05<br>(2.17) | 55.54<br>(2.78) | 168.97<br>(3.42) | 346.95<br>(2.61) | 864.69<br>(4.02) | 965.88<br>(2.01) |
| Meghalaya | 10.26<br>(1.85) | 29.72<br>(1.49) | 40.35<br>(0.82) | 71.69<br>(0.54) | 100.91<br>(0.47) | 56.78<br>(0.12) |
| Nagaland | 0.45<br>(0.08) | 2.66<br>(0.13) | 19.56<br>(0.40) | 17.62<br>(0.13) | 11.36<br>(0.05) | 11.40<br>(0.02) |
| Mizoram | 0.58<br>(0.10) | 2.66<br>(0.13) | 41.50<br>(0.84) | 18.25<br>(0.14) | 125.97<br>(0.59) | 203.24<br>(0.42) |
| Sikkim | 4.06<br>(0.73) | 18.81<br>(0.94) | 37.24<br>(0.75) | 26.94<br>(0.20) | 70.35<br>(0.33) | 26.42<br>(0.05) |
| Arunachal Pradesh | 0.37<br>(0.07) | 2.54<br>(0.13) | 6.64<br>(0.13) | 11.69<br>(0.09) | 18.99<br>(0.09) | 31.06<br>(0.06) |

*Source*: (1) Government of India, Fertilizers Association, Fertilizer Statistics, various issues, New Delhi.

(2) Government of India, Economic survey, union Budget, various years.

(3) Government of Punjab, Statistical Abstract, Various years.

*Note*: (1) Fertilizers subsidies per hectare of states are calculated by multiplying the consumption of fertilizers (in kgs) per hectare with subsidy per kg.

(2) Percentages are shown in parentheses.

In all the states of east zone, the fertilizers subsidies have increased in absolute terms during pre as well as post-liberalisation periods. In Odisha, these have increased the maximum *i.e.* near about fifteen times, in West Bengal near about nine times and in Bihar near about eight times in the year 2006-07 as compared to the year 1990-91. West Bengal has got near about two times more of fertilizers subsidies than that of Bihar during pre as well as post-liberalisation periods, whereas Bihar has received near about three times more of fertilizers subsidies and 1.49 times more of fertilizers subsidies in 1990-91 and 2006-07 respectively as compared to Odisha.

The fertilizers subsidies per hectare of north-east zone of India during 1980-81 to 2006-07 are shown in Table 4.19. It is found that in this zone, the subsidies have increased in all the states except in Nagaland, Mizoram and Sikkim during pre as well as post-liberalisation periods. Manipur has got topmost rank during the study period, in this state these have risen up by 360.74 per cent, 204.22 per cent, 105.34 per cent, 149.23 per cent and 11.70 per cent in 1985-86, 1990-91, 1996-97, 2000-01 and 2006-07 respectively.

In Assam, these have increased from Rs.2.37 in 1980-81 to Rs.36.71 in 1990-91 and further increased to Rs.670.72 in 2006-07. In Tripura, subsidies have gone up from Rs.4.65 in 1980-81 to Rs.76.66 in 1990-91 and further gone up to Rs.624.25 in 2006-07. On the other hand the percentage share of Meghalaya has declined during pre as well as post-liberalisation periods. The percentage share has increased from 0.08 in 1980-81 to 0.13 in 1985-86 and declined to 0.13 in 1996-97 and further declined to 0.02 in 2006-07 in Meghalaya.

In Mizoram, the percentage share of these subsidies have increased from 0.10 in 1980-81 to 0.84 in 1990-91 and declined to 0.59 in 2000-01 and further declined to 0.42 in 2006-07, whereas in Sikkim, it has increased from 0.73 per cent in 1980-81 to 0.94 per cent in 1985-86 and declined to 0.75 per cent in 1990-91 and further declined to 0.05 per cent in 2006-07.

The analysis of fertilizers subsidies per hectare shows that in all the states of north-east zone, these have increased except in Meghalaya, Nagaland, Mizoram and Sikkim in absolute terms during the study period. As the year 2006-07 is compared to the year 1990-91, it is observed that in Assam, these have risen up by eighteen times, in Tripura more than eight times, in Manipur near about six times, in Mizoram as well as in Arunachal Pradesh near about five and Meghalaya only 1.4 times. In 1990-91, Manipur has got near about five times more of fertilizers subsidies than that of Assam, Meghalaya has got more than six times than that of Arunachal Pradesh and Mizoram has received more than two times than that of Nagaland and in 2006-07, these states have got 1.44 times, near about two times, near about eighteen times more of fertilizers subsidies respectively.

## Section – III

India's electricity-supply industry is mainly owned and operated by the public sector. It is currently running a growing risk of bankruptcy. This has created a serious impediment to investments in the sector at a time when India desperately needs them. This is reflected in the sharp decrease of the ratio of electricity consumption growth to Gross Domestic Product (GDP) growth in the 1990s. In other words, in the past decade, electricity consumption growth did not follow economic growth. The State Electricity Boards (SEBs) end-use electricity tariffs vary widely according to customer category. The major categories are households, agriculture, commercial activities, industry and railways. There is large cross-subsidy between customer categories in India: tariffs for households and agriculture are generally well below actual supply costs, while tariffs to other customer categories are usually above the utilities' reported average cost of supply. Most of this subsidy is reported to be for the agricultural sector (Pierre, 2002).

Prior to 1948, private entities and local authorities generated approximately 80 per cent of electricity in India (Dubash, 2001). With the Electricity Supply Act of 1948, states gained control over electricity generation and each state organized a vertically integrated State Electricity Board (SEB). Though jurisdiction over electricity is percentage shared between the central and state governments, SEBs function as autonomous institutions. They have the authority to set and collect electricity tariffs, and are responsible for power generation and distribution. While SEBs has the authority to price electricity, electricity pricing has often been at the discretion of the state government and politicians rather than the SEBs (Gulati, 2003).

When electricity subsidy was first introduced, little competition existed between political parties for power. As agricultural profits increased, due in part to electricity subsidy, the farming workforce organized into a powerful political coalition. Around the same time, competition began to emerge between political parties for power. To gain the vote of the farmers, especially the well-organized landed class of farmers, politicians campaigned on at the rate of tariffs as opposed to metered tariffs. By 1989, the government was spending 25 per cent of total expenditure on agricultural electricity subsidy, and politicians were required to maintain this subsidy to either gain election or remain in power. For example in 2004, the Congress Party on Andhra Pradesh campaigned on free power; following its election to power, the party provided electricity free of charge to farmers (Dubash, 2007).

In India, commercial and industrial electricity users partially subsidize rural electricity users. These users often attribute the poor, sporadic and unreliable provision of electricity to the low prices paid by rural users (McKenzie and Ray 2004). In states where the commercial and industrial sectors comprise the inertial and dominant political lobbies, politicians follow a reverse campaign strategy. Politicians promise to reduce or eliminate the electricity subsidy provided to farmers (Badiani, 2010).

The subsidy of electricity of five zones in India during 1980-81 to 2008-09 is shown in Table 4.20. This table reveals that in India, power subsidy has increased throughout the study period, whereas declined in south zone, west zone and east zone in 2008-09. On the other hand, increased in north as well as north-east zones during pre and post-liberalisation periods.

In India, this subsidy has increased from Rs.357.56 crores in 1980-81 to Rs.1,324.15 in 1985-86 and further increased to Rs.26,904 crores in 2000-01 and declined to Rs.14,771.52 crores in 2008-09. In south zone, electricity subsidy has risen up from Rs.83.52 crores in 1980-81 to Rs.8,124 crores in 2000-01 and declined to Rs.4,071.13 crores in 2008-09. In west zone, power subsidy has gone up from Rs.83.96 crores in 1980-81 to Rs.1,588.17 crores in 1990-91 and further gone up to Rs.12,353 crores in 2000-01 and declined to Rs.3,209.67 crores in 2008-09. In east zone, this has increased from Rs.20.51 crores in 1980-81 to Rs.976 crores in 2000-01 and declined to Rs.720 crores in 2008-09. The electricity subsidy has increased from Rs.0.31 crores in 1980-81 to Rs.5.19 crores in 1990-91 and further increased to Rs.14 crores in 2000-01 in north-east zone.

South zone has got 23.36 percentage share at national level in 1980-81, which has increased to 31.05 in 1985-86 and declined to 27.68 in 1990-91 and again increased to 30.20 in 2000-01, whereas the percentage share of electricity subsidy has increased from 23.48 in 1980-81 to 34.37 in 1990-91 and further increased to 46.18 in 1996-97 and declined to 21.73 in 2008-09 in west zone.

North zone has got first rank in 1980-81 by receiving a major amount of percentage share (47.34 per cent) of electricity subsidy and this percentage share has declined to 31.29 in 1990-91 and increased to 45.84 in 2008-09. East zone has got fourth and north-east zone has occupied fifth position in case of electricity subsidy during pre as well as post-liberalisation periods 1980-81 to 2008-09. East zone has received 5.74 per cent, 7.77 per cent, 6.55 per cent, 4.21 per cent, 3.63 per cent and 4.87 per cent in 1980-81, 1985-86, 1990-91, 1996-97, 2000-01 and 2008-09 respectively, whereas it has declined from 0.09 per cent in 1980-81 to 0.08 per cent in 1985-86 and increased to 0.11 per cent in 1990-91 and again declined to 0.05 per cent in 2000-01 in north-east zone.

**Table 4.20: Zone-wise Distribution of Electricity Subsidy in India during 1980-18 to 2008-09 (In Rs. Crores)**

| *Years/Zones* | *1980-81* | *1985-86* | *1990-91* | *1996-97* | *2000-01* | *2008-09* |
|---|---|---|---|---|---|---|
| South | 83.52<br>(23.36) | 411.12<br>(31.05) | 1,278.95<br>(27.68) | 4,095<br>(26.26) | 8,124<br>(30.20) | 4,071.13<br>(27.56) |
| West | 83.96<br>(23.48) | 386.51<br>(29.19) | 1,588.17<br>(34.37) | 7,201<br>(46.18) | 12,353<br>(45.92) | 3,209.67<br>(21.73) |
| North | 169.26<br>(47.34) | 422.51<br>(31.91) | 1,445.98<br>(31.29) | 3,635<br>(23.31) | 5,437<br>(20.21) | 6,770.72<br>(45.84) |
| East | 20.51<br>(5.74) | 102.93<br>(7.77) | 302.71<br>(6.55) | 656<br>(4.21) | 976<br>(3.63) | 720<br>(4.87) |
| North-East | 0.31<br>(0.09) | 1.08<br>(0.08) | 5.19<br>(0.11) | 7<br>(0.04) | 14<br>(0.05) | – |
| India | 357.56<br>(100) | 1,324.15<br>(100) | 4,621<br>(100) | 15,594<br>(100) | 26,904<br>(100) | 14,771.52<br>(100) |

*Source*: Government of India, State Electricity Boards (SEBs), Annual Reports, various years.

*Note*: Percentages are shown in parentheses.

Above table shows that at India level, the electricity subsidy in Rs. crores has increased in absolute terms during the study period except in 2008-09, whereas the same pattern is found in south, west and east zones. As the year 2008-09 is compared to the year 1990-91, in north zone, this subsidy has increased by 4.68 times more of electricity subsidy, in south zone 3.18 three times, in east zone 2.37 times and in west zone 2.02 times. In 1990-91, the north zone has got 1.13 times more of electricity subsidy and 1.66 times in 2008-09 as compared to south zone. As compared to east zone, west zone has received 5.25 times more of electricity subsidy in 1990-91 and 4.46 times in 2008-09.

The electricity subsidy of different states of south zone in India during 1980-81 to 2008-09 is shown in Table 4.21. In Andhra Pradesh, the electricity subsidy has increased from Rs.16.5 crores in 1980-81 to Rs.1,571 crores in 1996-97 and further increased to Rs.3,685 crores in 2000-01. This has gone up from Rs.1.48 crores in 1980-81 to Rs.1,255 crores in 1996-97 and further gone up to Rs.1,899 crores in 2000-01 and declined to Rs.1,490.35 crores in 2008-09 in Karnataka.

**Table 4.21: State-wise Distribution of Electricity Subsidy in South Zone in India during 1980-81 to 2008-09 (In Rs. Crores)**

| *Years/States* | *South Zone* | | | | | |
|---|---|---|---|---|---|---|
| | *1980-81* | *1985-86* | *1990-91* | *1996-97* | *2000-01* | *2008-09* |
| Andhra Pradesh | 16.5<br>(4.61) | 124.01<br>(9.37) | 481.04<br>(10.41) | 1,571<br>(10.07) | 3,685<br>(13.70) | – |
| Karnataka | 1.48<br>(0.41) | 103.6<br>(7.82) | 332.77<br>(7.20) | 1,255<br>(8.05) | 1,899<br>(7.06) | 1,490.35<br>(10.09) |
| Kerala | 0.58<br>(0.16) | 0.63<br>(0.05) | 8.89<br>(0.19) | 44<br>(0.28) | 97<br>(0.36) | 749.17<br>(5.07) |
| Tamil Nadu | 64.96<br>(18.17) | 182.88<br>(13.81) | 456.25<br>(9.87) | 1,225<br>(7.86) | 2,443<br>(9.08) | 1,831.61<br>(12.40) |

*Source*: Government of India, State Electricity Boards (SEBs), Annual Reports, various years.

*Note*: Percentage is shown in parentheses.

It is found that in Kerala, this subsidy has increased during pre and post-liberalisation periods, whereas in Tamil Nadu, this has increased during pre as well as post-liberalisation periods except in 2008-09. This subsidy has increased from Rs.0.58 crores in 1980-81 to Rs.44 crores in 1996-97 and further increased to Rs.749.17 crores in 2008-09 in Kerala. The same subsidy has gone up from Rs.64.96 in 1980-81 to Rs.456.25 crores in 1990-91 and further gone up to Rs.2,443 crores in 2000-01 and declined to Rs.1,831.61 crores in 2008-09 in Tamil Nadu.

Tamil Nadu has got first rank by consuming maximum amount of electricity subsidy during 1980-81 to 1985-86, whereas Andhra Pradesh has got topmost position during 1990-91 to 2000-01. The percentage share of this subsidy in Andhra Pradesh, has increased to 10.41 in 1990-91 and declined to 10.07 in 1996-97 and increased to 13.70 in 2000-01, whereas the percentage share of electricity subsidy has increased from 0.41 in 1980-81 to 8.05 in 1996-97 and declined to 7.06 in 2000-01 in Karnataka state.

In Kerala, the percentage share has declined from 0.16 in 1980-81 to 0.05 in 1985-86 and increased to 0.28 in 1996-97 and further increased to 5.07 in 2008-09, on the other hand it has declined from 18.17 per cent in 1980-81 to 9.87 per cent in

1990-91 and further declined to 7.86 per cent in 1996-97 and increased to 12.40 per cent in 2008-09 in Tamil Nadu.

It is observed that in absolute terms, the electricity subsidy in Rs. crores has increased in all the states of south zones during 1980-81 to 2000-01. While comparing the year 2008-09 with the year 1990-91, in Kerala, this has increased by 84.27 times, in Karnataka 4.48 times and in Tamil Nadu 4.01 times more of electricity subsidy. As compared to Kerala, in 1990-91, Karnataka has got 37.43 times more of electricity subsidy, Tamil Nadu has got more than fifth one times, on the other hand, in 2008-09, in Karnataka, this has risen up by near about two times and in Tamil Nadu more than two times.

**Table 4.22: State-wise Distribution of Electricity Subsidy in West Zone in India during 1980-81 to 2008-09 (In Rs. Crores)**

| *Years/States* | *West Zone* | | | | | |
|---|---|---|---|---|---|---|
| | *1980-81* | *1985-86* | *1990-91* | *1996-97* | *2000-01* | *2008-09* |
| Gujarat | 24.12 | 66.81 | 539.27 | 1887 | 4577 | 1100 |
| | (6.75) | (5.05) | (11.67) | (12.10) | (17.01) | (7.45) |
| Madhya Pradesh | 6.03 | 33.65 | 245.6 | 1,795 | 3,134 | 906.34 |
| | (1.69) | (2.54) | (5.31) | (11.51) | (11.65) | (6.14) |
| Maharashtra | 33.78 | 211 | 591.09 | 2,556 | 2,586 | – |
| | (9.45) | (15.93) | (12.79) | (16.39) | (9.61) | |
| Rajasthan | 20.03 | 75.05 | 212.21 | 963 | 2,056 | 1,203.33 |
| | (5.60) | (5.67) | (4.59) | (6.18) | (7.64) | (8.15) |

*Source*: Government of India, State Electricity Boards (SEBs), Annual Reports, various years.

*Note*: Percentages are shown in parentheses.

The electricity subsidy of west zone during 1980-81 to 2008-09 is shown in Table 4.22. It is observed that in all the states of west zone, the power subsidy has increased in all the years except in 2008-09. Maharashtra is ahead among all the other states like Gujarat, Madhya Pradesh and Rajasthan during pre and post-liberalisation periods except in 2000-01. In Maharashtra, electricity subsidy has increased from Rs.33.78 crores in 1980-81 to Rs.591.09 crores in 1990-91 and further increased to Rs.2,586 crores in 2000-01. Gujarat has occupied second rank during 1980-81 to 1996-97 and first in 2000-01. This subsidy has gone up from Rs.24.12 crores in 1980-81 to Rs.66.81 crores in 1985-86 and further gone up to Rs.4,577.00 crores in 2000-01 and declined to Rs.1,100.00 crores in 2008-09 in Gujarat.

In Madhya Pradesh, electricity subsidy has increased from Rs.6.03 crores in 1980-81 to Rs.245.6 crores in 1990-91 and further increased to Rs.3,134 crores in 2000-01 and declined to Rs.906.34 crores in 2008-09, whereas this has increased from Rs.20.03 in 1980-81 to Rs.212.21 crores in 1990-91 and further increased to Rs.963 crores in 1996-97 and declined to Rs.1,203.33 crores in 2008-09 in Rajasthan.

The percentage share of Gujarat has increased from 6.75 in 1980-81 to 12.10 in 1996-97 and further increased to 17.01 in 2000-01 and declined to 7.45 in 2008-09, whereas the percentage share of Madhya Pradesh has risen up from 1.69 in 1980-81 to 5.31 in 1990-91 and further risen up to 11.51 in 1996-97 and declined to 6.14 in 2008-09. The percentage share of Rajasthan has increased during pre as well as post-liberalisation periods except in 1990-91. Maharashtra is the third state of west zone obtained 9.45 per cent, 12.79 per cent and 9.61 per cent in 1980-81, 1990-91 and 2000-01 respectively.

It is observed that in all the states of west zone, the electricity subsidy in Rs. crores has increased in absolute terms during pre as well as post-liberalisation periods except in 2008-09. As post-liberalisation period (2008-09) is compared to pre-liberalisation period (1990-91), in Rajasthan, this subsidy has increased near about six times, in Madhya Pradesh more than four times and in Gujarat only two times. Gujarat has got near about three times more of electricity subsidy and 1.2 times in 1990-91 and 2008-09 respectively as compared to Madhya Pradesh.

State-wise distribution of electricity subsidy in north zone of India during 1980-81 to 2008-09 in Table 4.23. The subsidy has increased in all the states of this zone except in 1996-97 and 2000-01. In Haryana, this has increased from Rs.22.57 crores in 1980-81 to Rs.770.00 crores in 1996-97 and further increased to Rs.2,636.99 crores in 2008-09. The subsidy of power has risen up from Rs.40.44 crores in 1980-81 to Rs.2,251 crores in 2000-01 and further rose to Rs.2,601.73 crores in 2008-09 in Punjab.

**Table 4.23: State-wise Distribution of Electricity Subsidy in North Zone in India during 1980-81 2008-09 (In Rs. Crores)**

| *Years/States* | *North Zone* | | | | | |
|---|---|---|---|---|---|---|
| | *1980-81* | *1985-86* | *1990-91* | *1996-97* | *2000-01* | *2008-09* |
| Haryana | 22.57<br>(6.31) | 70.44<br>(5.32) | 225.58<br>(4.88) | 770<br>(4.94) | 1,896<br>(7.05) | 2,636.99<br>(17.85) |
| Punjab | 40.44<br>(11.31) | 135.82<br>(10.26) | 499.66<br>(10.81) | 1,009<br>(6.47) | 2,251<br>(8.37) | 2,601.73<br>(17.61) |
| Uttar Pradesh | 105.46<br>(29.49) | 214.97<br>(16.23) | 702.86<br>(15.21) | 1762<br>(11.30) | 1257<br>(4.67) | 1,532<br>(10.37) |
| Jammu & Kashmir | – | – | 16<br>(0.35) | 93<br>(0.60) | 30<br>(0.11) | – |
| Himachal Pradesh | 0.79<br>(0.22) | 1.28<br>(0.10) | 1.88<br>(0.04) | 1<br>(0.01) | 3<br>(0.01) | – |

*Source*: Government of India, State Electricity Boards (SEBs), Annual Reports, various years.

*Note*: Percentages are shown in parentheses.

In Uttar Pradesh, this subsidy has gone up by 103.84 per cent in 1985-86 and 226.96 per cent in 1990-91 and 150.69 per cent in 1996-97 and declined by 28.66 per cent in 2000-01 and gone up by 21.88 per cent in 2008-09 as compared predecessor time given in table. In Jammu & Kashmir this has risen up from Rs.16 crores in 1990-91 to Rs.93 crores in 1996-97 and declined to Rs.30 crores in 2000-01. Variations are seen in Himachal Pradesh as electricity subsidy has increased from Rs.0.79 crores in 1980-81 to Rs.1.88 crores in 1990-91 and declined to Rs.1.00 crores in 1996-97 and again increased to Rs.3 crores in 2000-01.

Percentage wise analysis reveals that Uttar Pradesh has got first rank during 1980-81 to 1996-97, whereas it has lost its same position during 2000-01 to 2008-09. It is observed that in Jammu & Kashmir, the percentage share of power subsidy has increased except in 2000-01, whereas in Himachal Pradesh, it has declined throughout the study period. During 1980-81 to 2008-09, a lot of variation is found in the remaining three states (Haryana, Uttar Pradesh and Punjab).

The percentage share of Punjab has declined from 11.31 in 1980-81 to 6.47 in 1996-97 and increased to 8.37 in 2000-01 and further increased to 17.61 in 2008-09. Uttar Pradesh has got 29.49, per cent 15.21, 11.30 per cent and 10.37 per cent in 1980-81, 1990-91, 1996-97and 2008-09. On the other hand Himachal Pradesh has got 0.22 per cent, 0.04 per cent and 0.01 per cent in 1980-81, 1990-91 and 2000-01 respectively.

It is observed that in Haryana as well as in Punjab, electricity subsidy in Rs. crores has increased in absolute terms during pre and post-liberalisation periods, whereas in Uttar Pradesh and in Jammu & Kashmir, this subsidy has declined in 2000-01. As the year 2008-09 is compared to the year 1990-91, it is found that in Haryana, power subsidy has increased by near about twelve times more, in Punjab more than five and in Uttar Pradesh more than two times. In Punjab, this has increased more than two times in 1990-91 as compared to Haryana, whereas in 2008-09, Haryana has got 1.01 times more of electricity subsidy as compared to Punjab.

The electricity subsidy in east zone of India during 1980-81 to 2008-09 is shown in Table 4.24. In all the states of this zone the power subsidy has increased during the study period at different increasing rate throughout the study period. In Bihar, this subsidy has increased from Rs.18.36 crores in 1980-81 to Rs.351 crores in 1996-97 and further increased to Rs.720 crores in 2008-09, whereas this subsidy has increased from Rs.0.95 crores in 1980-81 Rs.12.35 crores in 1990-91 and further increased to Rs.48 crores in 1996-97 in Odisha.

**Table 4.24: State-wise Distribution of Electricity Subsidy in East Zone in India during 1980-81 to 2008-09 (In Rs. Crores)**

| *Years/States* | *East Zone* | | | | | |
|---|---|---|---|---|---|---|
| | *1980-81* | *1985-86* | *1990-91* | *1996-97* | *2000-01* | *2008-09* |
| Bihar | 18.36<br>(5.13) | 92.21<br>(6.96) | 234.08<br>(5.07) | 351<br>(2.25) | 586<br>(2.18) | 720<br>(4.87) |
| Odisha | 0.95<br>(0.27) | 2.48<br>(0.19) | 12.35<br>(0.27) | 48<br>(0.31) | – | – |
| West Bengal | 1.2<br>(0.34) | 8.24<br>(0.62) | 56.28<br>(1.22) | 257<br>(1.65) | – | – |

*Source*: Government of India, State Electricity Boards (SEBs), Annual Reports, various years.

*Note*: Percentages are shown in parentheses.

In West Bengal, this subsidy has increased from Rs.1.2 crores in 1980-81 to Rs.56.28 crores in 1990-91 and further increased to Rs.390 crores in 2000-01. The percentage share at country level has increased from 5.13 in 1980-81 to 6.96 in 1985-86 and declined to 5.07 in 1990-91 and further declined to 2.18 in 2000-01. The percentage share of Odisha has declined from 0.27 in 1980-81 to 0.31 in 1996-97, whereas 0.34 per cent, 1.22 per cent, 1.65 per cent and 1.45 per cent in 1980-81, 1990-91, 1996-97, 2000-01 and 2008-09 is received by West Bengal.

The above table concluded that in absolute terms, the electricity subsidy in Rs. crores has increased in all the states of east zone throughout the study period. As post-liberalisation period (1996-97) is compared to pre liberalisation period (1980-81), it is found that in West Bengal this has risen up by 214.17 times, in Odisha 50.53 and in Bihar 19.12 times.

The electricity subsidy in north-east zone of India during 1980-81 to 2008-09 is shown in Table 4.25. This table shows that in Assam, electricity subsidy has gone up from Rs.0.31 crores in 1980-81 to Rs.5.19 crores in 1990-91 and further gone up 14 crores in 2000-01. The percentage share of electricity subsidy has declined from 0.09 in 1980-81 to 0.11 in 1990-91 and further declined to 0.04 in 1996-97 and increased to 0.05 in 2000-01 in Assam.

**Table 4.25: State-wise Distribution of Electricity Subsidy in North-East Zone in India during 1980-81 to 2008-09 (In Rs. Crores)**

| *Years/States* | *North-East Zone* | | | | | |
|---|---|---|---|---|---|---|
| | *1980-81* | *1985-86* | *1990-91* | *1996-97* | *2000-01* | *2008-09* |
| Assam | 0.31<br>(0.09) | 1.08<br>(0.08) | 5.19<br>(0.11) | 7<br>(0.04) | 14<br>(0.05) | – |

*Source*: Government of India, State Electricity Boards (SEBs), Annual Reports, various years.

*Note*: (1) Data is available for Assam only in reports, other states are not discussed.

(2) Percentages are shown in parentheses..

It is found that in this zone, the electricity subsidy in Rs. crores has increased during pre and post-liberalisation periods, whereas the increasing rate is higher in 1990-91 among all the other years of study. The increasing rate is 248.39 per cent in 1985-86, 380.56 per cent in 1990-91, 34.87 per cent in 1996-97 and 14 per cent in 2000-01. As 2000-01 is compared to 1980-81, it is observed that in Assam this has increased by more than forty five times.

The distribution of electricity subsidy per hectare in five zones of India during 1980-81 to 2000-01 is shown in Table 4.26. Considering that gross cropped area varies across the states, electricity subsidy per hectare basis has been calculated. It is observed that per hectare subsidy of electricity has increased in India and in all the zones during pre as well as post-liberalisation periods. In 1980-81 and in 1990-91, north zone has got first rank, whereas in 1985-86, 1996-97 and 2000-01, south zone ranked first.

In India, per hectare subsidy of electricity has increased from Rs.343.26 in 1980-81 to Rs.24,472.64 in 2000-01. In south zone, this has increased from Rs.117.27 in 1980-81 to Rs.4,231.81 in 1996-97 and further increased to Rs.8,441.98 in 2000-01, whereas in west zone this subsidy gone up from Rs.53.58 in 1980-81 to Rs.8,284.74 in 2000-01. North zone has got Rs.152.38 in 1980-81, increased by 171.92 per cent,

260.39 per cent, 174.75 per cent and 63.32 per cent in 1985-86, 1990-91 and 1996-97 respectively.

In east zone, this subsidy has increased from Rs.19.13 in 1980-81 to Rs.678.50 in 1996-97 and further increased to Rs.1,010.97 in 2000-01. In north-east zone, this has increased from Rs.0.90 in 1980-81 to Rs.17.58 in 1996-97 and further increased to Rs. 34.44 in 2000-01. North zone has occupied first rank by getting 44.39 percentage share of electricity subsidy per hectare followed by south zone (34.16 per cent), west zone (15.61 per cent), east zone (5.57 per cent) and north-east zone (0.26 per cent) in 1980-81, whereas in 1985-86, south zone achieved first position by receiving 38.45 per cent followed by north zone (34.20 per cent), west zone (18.80 per cent), east zone (8.31 per cent) and north-east zone (0.23 per cent).

In south zone, this subsidy has declined from Rs.33.43 in 1990-91 to 32.25 in 1996-97 and increased to Rs.34.50 in 2000-01 whereas west zone has enjoyed 22.30 per cent, 31.18 per cent and 33.85 per cent of electricity subsidy in 1990-91, 1996-97 and 2000-01 respectively. It is observed that percentage share of north zone as well as east has declined during 1990-91 to 2000-01. The percentage share of electricity subsidy has declined from 0.33 in 1980-81 to 0.13 in 1996-97 and increased to 0.14 in 2000-01 in north-east zone.

**Table 4.26: Zone-wise Electricity Subsidy in India during 1980-81 to 2000-01 (In Rs./Hectare)**

| *Years/Zones* | *1980-81* | *1985-86* | *1990-91* | *1996-97* | *2000-01* |
|---|---|---|---|---|---|
| South | 117.27 | 465.83 | 1,365.03 | 4,231.81 | 8,441.98 |
| | (34.16) | (38.45) | (33.43) | (32.25) | (34.50) |
| West | 53.58 | 227.74 | 910.72 | 4,091.48 | 8,284.74 |
| | (15.61) | (18.80) | (22.30) | (31.18) | (33.85) |
| North | 152.38 | 414.36 | 1,493.28 | 4,102.78 | 6,700.51 |
| | (44.39) | (34.20) | (36.57) | (31.27) | (27.38) |
| East | 19.13 | 100.67 | 301.10 | 678.50 | 1,010.97 |
| | (5.57) | (8.31) | (7.37) | (5.17) | (4.13) |
| North-East | 0.90 | 2.85 | 13.67 | 17.58 | 34.44 |
| | (0.26) | (0.23) | (0.33) | (0.13) | (0.14) |
| India | 343.26 | 1,211.44 | 4,083.80 | 13,122.15 | 24,472.64 |
| | (100) | (100) | (100) | (100) | (100) |

*Source*: (1) Government of India, State Electricity Boards (SEBs), Annual Reports, various years.

(2) Government of Punjab, Statistical Abstract, various years.

*Note*: (1) Electricity subsidies per hectare of zones are calculated by adding the electricity subsidies per hectares on zone basis.

(2) Percentages are shown in parentheses.

Above table shows that at national as well as zones level, the electricity subsidy per hectare has increased in absolute terms during pre as well as post-liberalisation periods. As the year 2000-01 is compared to the year 1990-91, it is found that in India, it has increased by near about six times more, in west zone nine times, south zone more than six times, in north zone near about four times, in east zone 3.4 times and north-east zone more than two times. North zone has got approximately five times more of electricity subsidy and near about seven times as compared to east in 1990-91 and 2000-01 respectively.

The state-wise electricity of subsidy in south zone of India during 1980-81 to 2000-01 is shown in Table 4.27. It is observed that in all the stares of south zone, the electricity subsidy has increased throughout the study period. In Andhra Pradesh, this has increased from Rs.13.44 in 1980-81 to Rs.1,171.51 in 1996-97 and further increased to Rs.2,720.56 in 2000-01, whereas in Karnataka, electricity subsidy has risen up from Rs.1.39 in 1980-81 to Rs.1,545.91 in 2000-01. Kerala has got Rs.2.03 in 1980-81, increased to Rs.29.44 in 1990-91 and further increased to Rs.320.98 in 2000-01, on the other hand, this subsidy has increased from Rs.100.42 in 1980-81 to Rs.687.95 in 1990-91 and further increased to Rs. 3,854.53 in 2000-01 in Tamil Nadu.

This table indicates that in Andhra Pradesh, the percentage share of subsidy at India level has increased, whereas in Karnataka, Kerala and in Tamil Nadu a lot of variation is seen during the study period. It is observed that in Andhra Pradesh the percentage share has increased from 3.91 in 1980-81 to 8.93 in 1996-97 and further increased to 11.12 in 2000-01. In Karnataka a lot of variation is seen in percentages-wise analysis.

In Kerala, the percentage share has declined from 0.59 in 1980-81 to 0.18 in 1985-86 and gone up to 0.72 in 1990-91 and further gone up to 1.31 in 2000-01, on the other hand it has declined from 29.25 in 1980-81 to 16.85 in 1990-91 and further declined to 14.46 in 1996-97 and increased to 15.75 in 2000-01 in Tamil Nadu.

**Table 4.27: State-wise Electricity Subsidy in South Zone in India during 1980-81 to 2000-01(In Rs./Hectare)**

| *Years/States* | *South Zone* | | | | |
|---|---|---|---|---|---|
| | *1980-81* | *1985-86* | *1990-91* | *1996-97* | *2000-01* |
| Andhra Pradesh | 13.44<br>(3.91) | 102.49<br>(8.46) | 364.65<br>(8.93) | 1171.51<br>(8.93) | 2720.56<br>(11.12) |
| Karnataka | 1.39<br>(0.40) | 92.95<br>(7.67) | 282.99<br>(6.93) | 1017.43<br>(7.75) | 1545.91<br>(6.32) |
| Kerala | 2.03<br>(0.59) | 2.20<br>(0.18) | 29.44<br>(0.72) | 145.70<br>(1.11) | 320.98<br>(1.31) |
| Tamil Nadu | 100.42<br>(29.25) | 268.19<br>(22.14) | 687.95<br>(16.85) | 1897.17<br>(14.46) | 3854.53<br>(15.75) |

*Source*: (1) Government of India, State Electricity Boards (SEBs), Annual Reports, various years.

(2) Government of Punjab, Statistical Abstract, various years.

*Note*: (1) Electricity subsidies per hectare of states are calculated by dividing the electricity subsidies in Rs. Crores with gross cropped area of the concerned state.

(2) Percentages are shown in parentheses.

It is observed that in absolute terms, the electricity subsidy per hectare has increased in all the states of south zone during the study period. As the year 2000-01 is compared to the year 1990-91, the above table indicates that in Kerala, this has increased the maximum times *i.e.* near about eleven, in Andhra Pradesh more than seven times, Tamil Nadu as well as in Karnataka near about six times. Tamil Nadu has got 2.4 times more of electricity subsidy as compared to Karnataka in 1990-91, whereas same pattern is also observed in post-liberalisation period (2000-01). On the other hand, in Andhra Pradesh, this subsidy has increased 12.39 times more and 8.48 times as compared to Kerala in 1990-91 and 2000-01 respectively.

The electricity subsidy of four states of west zone in India during 1980-81 to 2000-01 is shown in Table 4.28. It is found that electricity subsidy per hectare has increased in all the states except in Madhya Pradesh (in this stare this subsidy has declined in 1990-91).

**Table 4.28: State-wise Electricity Subsidy in West Zone in India during 1980-81 to 2000-01 (In Rs./Hectare)**

| *Years/States* | *West Zone* | | | | |
|---|---|---|---|---|---|
| | *1980-81* | *1985-86* | *1990-91* | *1996-97* | *2000-01* |
| Gujarat | 22.55<br>(6.57) | 69.00<br>(5.70) | 520.48<br>(12.75) | 1,715.30<br>(13.07) | 4,281.57<br>(17.50) |
| Madhya Pradesh | 2.82<br>(0.82) | 14.62<br>(1.21) | 10.28<br>(0.25) | 734.12<br>(5.59) | 1,753.78<br>(7.17) |
| Maharashtra | 16.67<br>(4.85) | 102.74<br>(8.48) | 270.46<br>(6.62) | 1,176.69<br>(8.97) | 1,180.23<br>(4.82) |
| Rajasthan | 11.54<br>(3.36) | 41.38<br>(3.42) | 109.50<br>(2.68) | 465.37<br>(3.55) | 1,069.16<br>(4.37) |

*Source*: (1) Government of India, State Electricity Boards (SEBs), Annual Reports, various years.

(2) Government of Punjab, Statistical Abstract, various years.

*Note*: (1) Electricity subsidies per hectare of states are calculated by dividing the electricity subsidies in Rs. Crores with gross cropped area of the concerned state.

(2) Percentages are shown in parentheses.

This table reveals that in Gujarat, this has increased from Rs.22.55 in 1980-81 to Rs.1,715.30 in 1996-97 and further increased to Rs.4,281.57 in 2000-01. This has increased from Rs.2.82 in 1980-81 to Rs.10.28 in 1990-91 and further increased to Rs.1,753.78 in 2000-01 in Madhya Pradesh.

In Maharashtra, this has increased from Rs.16.67 in 1980-81 to Rs.270.46 in 1990-91 and further increased to Rs.1180.23 in 2000-01, on the other hand, in Rajasthan, this subsidy has risen up from Rs.11.54 in 1980-81 to Rs.465.37 in 1990-91 and further risen up to Rs.1,069.16 in 2000-01. The percentage share of Gujarat has declined from 22.55 in 1980-81 to 5.70 in 1985-86 and increased to 17.50 in 2000-01, whereas the percentage share of electricity subsidy has increased from 0.82 in 1980-81 to 5.59 in 1996-97 and further increased to 7.17 in 2000-01 in Madhya Pradesh.

It is found that in Maharashtra, the percentage share has risen up from 4.85 in 1980-81 to 6.62 in 1990-91 and further declined to 4.82 in 2000-01, on the other hand, it has increased from 3.36 per cent in 1980-81 to 3.42 per cent in 1985-86 and declined to 2.68 per cent in 1990-91 and again increased to 4.37 per cent in 2000-01 in Rajasthan.

In absolute terms, the electricity subsidy per hectare has increased in all the states of west zone during pre as well as post-liberalisation periods. As post-liberalisation period (2000-01) is compared with pre-liberalisation period (1990-91), it is observed that in Madhya Pradesh, this subsidy has increased by 170.60 times, Rajasthan 9.76 times, Gujarat 8.23 times and Maharashtra more than four times. In Gujarat, this subsidy has increased by more than fifty times and more than two times as compared to Madhya Pradesh in 1990-91 and 2000-01 respectively.

The electricity subsidy per hectare of five states of north zone of India during 1980-81 to 2000-01 is shown in Table 4.29. It is observed that in Haryana and in Punjab this subsidy has increased, whereas declined in Jammu & Kashmir, Uttar Pradesh (in 2000-01) and in Himachal Pradesh (in 1996-97).

**Table 4.29: State-wise Electricity Subsidy in North Zone in India during 1980-81 to 2000-01 (In Rs./Hectare)**

| *Years/States* | *North Zone* | | | | |
|---|---|---|---|---|---|
| | *1980-81* | *1985-86* | *1990-91* | *1996-97* | *2000-01* |
| Haryana | 41.32<br>(12.04) | 125.76<br>(10.38) | 381.11<br>(9.33) | 1,267.70<br>(9.66) | 3,100.57<br>(12.67) |
| Punjab | 59.80<br>(17.42) | 189.75<br>(15.66) | 666.04<br>(16.31) | 1,286.66<br>(9.81) | 2,834.66<br>(11.58) |
| Jammu & Kashmir | – | – | 150.09<br>(3.68) | 863.51<br>(6.58) | 269.06<br>(1.10) |
| Himachal Pradesh | 8.35<br>(2.43) | 13.14<br>(1.08) | 20.19<br>(0.49) | 10.56<br>(0.08) | 31.65<br>(0.13) |
| Uttar Pradesh | 42.92<br>(12.50) | 85.71<br>(7.08) | 275.85<br>(6.75) | 674.35<br>(5.14) | 464.57<br>(1.90) |

*Source*: (1) Government of India, State Electricity Boards (SEBs), Annual Reports, various years.

(2) Government of Punjab, Statistical Abstract, various years.

Note: (1) Electricity subsidies per hectare of states are calculated by dividing the electricity subsidies in Rs. Crores with gross cropped area of the concerned state.

(2) Percentages are shown in parentheses.

In Haryana, this has increased from Rs.41.32 in 1980-81 to Rs.1,267.70 in 1996-97 and further increased to Rs.3,100.57 in 2000-01. In Punjab, the electricity subsidy per hectare has risen up from Rs.59.80 in 1980-81 to Rs.1,286.66 in 1996-97 and further risen up to Rs.2,834.66 in 2000-01, whereas this has gone up by 475.31 per cent in 1996-97 from 1980-81 and declined by 68.84 per cent in 2000-01 from 1996-97 in Jammu & Kashmir.

In Himachal Pradesh, this has increased from Rs.8.35 in 1980-81 to Rs.20.19 in 1990-91 and declined to Rs.10.56 in 1996-97 and again increased to Rs.31.65 in 2000-01, on the other hand, electricity subsidy has increased from Rs.42.92 in 1980-81 to Rs.275.85 in 1990-91 and further increased to Rs.674.35 in 1996-97 and declined to Rs.464.57 in 2000-01 in Uttar Pradesh.

It is seen that Punjab is ahead among all the other states of the same zone in case of receiving electricity subsidy during 1980-81 to 1996-97, whereas it has lost its rank in 2000-01. The percentage share of Haryana has declined from 12.04 in 1980-81 to 9.33 in 1990-91 and increased to 12.67 in 2000-01. The percentage share of Punjab has gone up from 17.42 in 1980-81 to 16.31 in 1990-91 and declined to 9.81 in 1996-97 and again gone up to 11.58 in 2000-01, whereas in Jammu & Kashmir, it has increased from 3.68 per cent in 1990-91 to 1.10 per cent in 2000-01 and Himachal Pradesh's percentage share has declined from 2.43 per cent in 1980-81 to 0.49 per cent in 1990-91 and further declined to 0.08 per cent and increased to 0.13 per cent in 2000-01. The percentage share of electricity subsidy per hectare has declined from 12.50 in 1980-81 to 1.90 in 2000-01 in Uttar Pradesh.

From the above table, it is concluded that in all the states (except in Jammu & Kashmir and Himachal Pradesh) of north zone, the electricity subsidy per hectare has increased in absolute terms, whereas in case of percentage share a lot of variation is seen during pre as well as post-liberalisation periods. As the year 2000-01 is compared to the year 1990-91, in Haryana, this has increased by more than eight times, in Punjab 4.26 times, whereas in Jammu & Kashmir, Himachal Pradesh and in Uttar Pradesh this has increased by near about two times. In 1990-91 and in 2000-01, in Haryana this has increased by 2.54 times and 11.52 times as compared to Jammu & Kashmir, on the other hand Punjab has got 2.41 times and 6.1 times as compared to Uttar Pradesh.

**Table 4.30: State-wise Electricity Subsidy in East Zone in India during 1980-81 to 2000-01 (In Rs./Hectare)**

| *Years/States* | *East Zone* | | | | |
|---|---|---|---|---|---|
| | *1980-81* | *1985-86* | *1990-91* | *1996-97* | 2000-01 |
| Bihar | 16.47<br>(4.80) | 87.68<br>(7.24) | 223.25<br>(5.47) | 346.12<br>(2.64) | 583.20<br>(2.38) |
| West Bengal | 1.57<br>(0.46) | 10.32<br>(0.85) | 64.97<br>(1.59) | 283.70<br>(2.16) | 427.77<br>(1.75) |
| Odisha | 1.09<br>(0.32) | 2.68<br>(0.22) | 12.87<br>(0.32) | 48.69<br>(0.37) | – |

*Source*: (1) Government of India, State Electricity Boards (SEBs), Annual Reports, various years.

(2) Government of Punjab, Statistical Abstract, various years.

Note: (1) Electricity subsidies per hectare of states are calculated by dividing the electricity subsidies in Rs. Crores with gross cropped area of the concerned state.

(2) Percentages are shown in parentheses.

The amount of electricity subsidy different states of east zone of India during 1980-81 to 2000-01 is shown in Table 4.30. This table reveals that in three states (Bihar, West Bengal and Odisha) of the same zone, the power subsidy has increased during pre as well as post-liberalisation periods.

In Bihar, this subsidy has increased from Rs.16.47 in 1980-81 to Rs.346.12 in 1996-97 and further increased to Rs.583.20 in 2000-01. In West Bengal, this has increased from Rs.1.57 in 1980-81 to Rs.283.70 in 1996-97 and further increased to 427.77 in 2000-01, whereas the same subsidy gone up from Rs.1.09 in 1980-81 to Rs.12.87 in 1990-91 and further gone up to Rs.48.69 in 1996-97 in Odisha.

The percentage-wise analysis indicates that Bihar is getting a maximum amount of electricity subsidy during the study period followed by West Bengal and Odisha. Bihar has received 4.80 per cent, 7.24 per cent, 5.47 per cent, 2.64 per cent and 2.38 per cent in 1980-81, 1985-86, 1990-91, 1996-97 and 2000-01 respectively. It is observed that the percentage share of West Bengal is increased during 1980-81 to 1996-97 and declined in 2000-01. The percentage share of in this state has increased from 0.46 in 1980-81 to 1.59 in 1990-91 and further increased to 2.16 and declined to 1.75 in 2000-01, whereas the percentage share of electricity subsidy per hectare has declined from 0.32 in 1980-81 to 0.22 in 1985-86 and increased to 0.32 in 1990-91 and further increased to 0.37 in 1996-97 in Odisha.

It is observed that in all the states of east zone, electricity subsidy per hectare has increased in absolute terms during pre as well as post-liberalisation periods. In Bihar, this subsidy has increased by 10.49 times, 3.43 times and 1.36 times as compared to West Bengal in 1980-81, in 1990-91 and in 2000-01 respectively. As the year 2000-01 is compared to the year 1990-91, Bihar has got 2.6 times more and West Bengal has got 6.5 times more of electricity subsidy.

**Table 4.31: State-wise Electricity Subsidy in North-East Zone in India during 1980-81 to 2000-01 (In Rs./Hectare)**

| *Years/States* | *North-East Zone* | | | | |
|---|---|---|---|---|---|
| | *1980-81* | *1985-86* | *1990-91* | *1996-97* | 2000-01 |
| Assam | 0.90<br>(0.26) | 2.85<br>(0.23) | 13.67<br>(0.33) | 17.58<br>(0.13) | 34.44<br>(0.14) |

*Source*: (1) Government of India, State Electricity Boards (SEBs), Annual Reports, various years.

(2) Government of Punjab, Statistical Abstract, various years.

*Note*: (1) Electricity subsidies per hectare of states are calculated by dividing the electricity subsidies in Rs. Crores with gross cropped area of the concerned state.

(2) Data is available for Assam only in reports, other states are not discussed.

(3) Percentages are shown in parentheses.

The per hectare subsidy of electricity of north-east zone of India during 1980-81 to 2000-01 is shown in Table 4.31. It table reveals that in Assam electricity subsidy per hectare has increased throughout the study period. The electricity subsidy per hectare of states is calculated by dividing the electricity subsidy in Rs. Crores with gross cropped area of the concerned state. In Assam, this type of subsidy has increased from Rs.0.90 in 1980-81 to Rs.13.67 in 1990-91 and further increased to Rs.34.44 in 2000-01. The percentage share of electricity subsidy has declined from 0.26 in 1980-81 to 0.13 in 1996-97 and increased to 0.14 in 2000-01 in Assam.

From the above table, it is observed that in absolute terms, electricity subsidy per hectare has increased during the study period. It is observed that in Assam, this subsidy has increased by 2.52 times in 2000-01 as compared to 1990-91. It is observed that the increasing rate is higher in 1990-91 among all the other years of study. In 1985-86, this subsidy has increased by 216.43 per cent, 380.18 per cent in 1990-91, 28.64 per cent in 1996-97 and 95.87 per cent in 2000-01 as compared predecessor times given in the table.

## Section – IV

In an agrarian economy like India, irrigation has played a major role in the agricultural production process. Irrigation development in the country has been taken up in a big way through Major, Medium and Minor irrigation schemes since independence. Agricultural prices, taxes and subsidy are powerful policy tools for a government that wants to advance its production or welfare goals. For its policies to be effective, it must predict how heterogeneous groups of farmers will alter their chosen crops and water use. A key determinant of a farmer's response is his or her access to water, especially where agriculture depends on irrigation from a canal. Other things being equal, downstream farmers in a canal suffer disproportionately from up- stream seepage. They also lose more water to theft, a widely observed phenomenon on canals in India and elsewhere. A government concerned about the welfare implications of existing prices should explicitly acknowledge that many farmers steal water. conversely, a government concerned about the social cost of water theft must understand the contribution of its own price distortions (Ashra, 2007).

The irrigation subsidy in five zones of India during 1980-81 to 2006-07 is shown in Table 4.32. It shows that in India as well as in the zones, this subsidy has increased at different increasing rate during pre as well as post-liberalisation periods.

In south zone, irrigation subsidy has risen up from Rs.139.03 crores in 1980-81 to Rs.15,223.37 crores in 2006-07, whereas in west zone, this has increased from Rs.129.68 crores in 1980-81 to Rs.12,338.08 crores in 2006-07. In north zone, this has gone up by 338.61 per cent in 1985-86,160.33 per cent in 1990-91,50.70 per cent in 1996-97,52.56 per cent in 2000-01 and 105.53 per cent in 2006-07 as compared predecessor years given in the table.

On the other hand, in east zone, subsidy of irrigation has increased from Rs.33.68 crores in 1980-81 to Rs.1861.20 crores in 2006-07. North-east zone has got Rs.8.66 crores, Rs.74.30 crores, Rs.74.40 crores and Rs.301.74 crores in 1980-81, 1990-91, 1996-97 and 2006-07 respectively.

It is seen that in 1980-81, south zone has got topmost position by getting huge amount of subsidy of irrigation at country level followed by west zone (32.49 per cent), north zone (22.06 per cent), east zone (8.44 per cent) and north-east zone (2.17

per cent). During 1985-86 to 2000-01, west zone has occupied first rank followed by south, north, east and north-east zones. Again in 2006-07, south zone achieved topmost rank, west, north, east and north-east zones has got second, third, fourth and fifth position.

In south zone, the percentage share has declined from 34.84 in 1980-81 to 44.16 in 2006-07, on the other hand, it has risen up from 32.49 per cent in 1980- 81 to 42.03 per cent in 1990-91 and further risen up to 48.04 per cent in 1996-97 and declined to 35.79 per cent in 2006-07 in west zone.

The percentage share of north zone increased from 22.06 in 1980-81 to 25.66 in 1990-91 and declined to 13.78 in 2006-07. In east zone, the percentage share has increased from 8.44 in 1980-81 to 9.81 in 1990-91 and declined to 8.88 in 2000-01 and further declined to 5.40 in 2006-07. North-east zone has got 2.17 per cent, 1.73 per cent, 1.90 per cent, 0.72 per cent and 0.88 per cent in 1980-81, 1985-86, 1990-91, 1996-97 and 2006-07 respectively.

**Table 4.32: Zone-wise Distribution of Irrigation Subsidy in India during 1980-81 to 2006-07 (In Rs. Crores)**

| *Years/Zones* | *1980-81* | *1985-86* | *1990-91* | *1996-97* | *2000-01* | *2006-07* |
|---|---|---|---|---|---|---|
| South | 139.03<br>(34.84) | 407.94<br>(24.47) | 806.78<br>(20.59) | 2,929.42<br>(28.15) | 4,721.18<br>(32.09) | 15,223.37<br>(44.16) |
| West | 129.68<br>(32.49) | 658.78<br>(39.51) | 1,646.68<br>(42.03) | 4,998.84<br>(48.04) | 6,266.48<br>(42.60) | 12,338.08<br>(35.79) |
| North | 88.05<br>(22.06) | 386.20<br>(23.16) | 1,005.39<br>(25.66) | 1,515.13<br>(14.56) | 2,311.48<br>(15.71) | 4,750.87<br>(13.78) |
| East | 33.68<br>(8.44) | 185.40<br>(11.12) | 384.26<br>(9.81) | 886.94<br>(8.52) | 1,306.28<br>(8.88) | 1,861.20<br>(5.40) |
| North-East | 8.66<br>(2.17) | 28.89<br>(1.73) | 74.30<br>(1.90) | 74.40<br>(0.72) | 106.29<br>(0.72) | 301.74<br>(0.88) |
| India | 399.10<br>(100) | 1,667.21<br>(100) | 3,917.41<br>(100) | 10,404.73<br>(100) | 14,711.71<br>(100) | 34,475.26<br>(100) |

*Source*: (1) Government o India, Pricing of Water in Public System in India 2010, Combined Finance and Revenue Accounts of different states.

*Note*: (1) Irrigation subsidies in Rs. Crores of zones are calculated by adding the irrigation subsidies on zone basis.

(2) Percentages are shown in parentheses.

It is observed that at national level as well as zone level, the irrigation subsidy in Rs. crores has increased in absolute terms during pre as well post-liberalisation periods. As post-liberalisation period (2006-07) is compared with post-liberalisation (1990-91), it is found at national level, this has increased by near about nine times, whereas in south zone this has increased by near about nineteen times, in west zone has got more than seven times, in east as well as in north-east zones near about five and in north zone more than four times of irrigation subsidy. In 1990-91, west zone has got more than two times and north zone near about three times more of irrigation subsidy as compared to south and east respectively, whereas in post-liberalisation period (2006-07), south zone has received 1.23 times and north zone near about three times more of irrigation subsidy as compared to west zone and east zone respectively.

The irrigation subsidy in south zone of India during 1980-81 to 2006-07 is shown in Table 4.33. This table reveals that in all the states the irrigation subsidy has increased, whereas Andhra Pradesh is leading among all the other states by getting maximum percentage share throughout the study period.

**Table 4.33: State-wise Distribution of Irrigation Subsidy in South Zone in India during 1980-81 to 2006-07 (In Rs. Crores)**

| *Years/States* | *South Zone* | | | | | |
|---|---|---|---|---|---|---|
| | *1980-81* | *1985-86* | *1990-91* | *1996-97* | *2000-01* | *2006-07* |
| Andhra Pradesh | 59.3<br>(14.86) | 188.4<br>(11.30) | 370.77<br>(9.46) | 1,305.09<br>(12.54) | 2,199.55<br>(14.95) | 10,557.2<br>(30.62) |
| Karnataka | 47.9<br>(12.00) | 140.73<br>(8.44) | 216.47<br>(5.53) | 1,249.47<br>(12.01) | 1,731.77<br>(11.77) | 3,792.45<br>(11.00) |
| Kerala | 6.94<br>(1.74) | 22.69<br>(1.36) | 82.23<br>(2.10) | 173.49<br>(1.67) | 179.86<br>(1.22) | 222.37<br>(0.65) |
| Tamil Nadu | 24.89<br>(6.24) | 56.12<br>(3.37) | 137.31<br>(3.51) | 200.82<br>(1.93) | 608.81<br>(4.14) | 651.35<br>(1.89) |
| Pondicherry | – | – | – | 0.55<br>(0.01) | 1.19<br>(0.01) | – |

*Source*: (1) Government of India, Pricing of Water in Public System, 2010, Combined Finance and Revenue Accounts of different states.

*Note*: (1) Irrigation subsidies in Rs. Crores of different states are calculated by difference between expenditure and revenue of medium, minor as well as major irrigations.

(2) Percentages are shown in parentheses.

In Andhra Pradesh, subsidy of irrigation has increased from Rs.59.3 crores in 1980-81 to Rs.10,557.20 crores in 2006-07, whereas in Karnataka, this has risen up from Rs.47.90 crores in 1980-81 to Rs.3,792.45 crores in 2006-07. Kerala has received Rs.6.94 crores in 1980-81, gone up by 226.95 per cent in 1985-86 and further gone up by 23.64 per cent in 2006-07. This has increased from Rs.24.89 crores in 1980-81 to Rs.608.81 crores in 2000-01 and further increased to Rs.651.35 crores in 2006-07 in Tamil Nadu.

The percentage share of Andhra Pradesh has declined from 14.86 in 1980-81 to 9.46 in 1990-91 and increased to 30.62 in 2006-07, whereas in Karnataka, the percentage share has declined from 12.00 in 1980-81 to 5.53 in 1990-91 and increased to 12.01 in 1996-97 and again declined to 11.00 in 2006-07. The percentage share of Kerala has declined from 1.74 in 1980-81 to 2.10 in 1990-91 and declined to 0.65 in 2006-07. In Tamil Nadu, it has declined from 6.24 per cent in 1980-81 to 1.93 per cent in 1996-97 and increased to 4.14 per cent in 2000-01. During 1996-97 to 2000-01, it is observed that the percentage share of Pondicherry (0.01 per cent) remains constant.

Above table reveals that in absolute terms, the irrigation subsidy in Rs. crores has increased in all the states of south zone during the study period. In Andhra Pradesh, this has increased the maximum times *i.e.* more than twenty eight, in Karnataka more than seventeen times, Tamil Nadu near about five times and in Kerala near about three times in 2006-07 as compared to 1990-91.

The irrigation subsidy in west zone of India during 1980-81 to 2006-07 is shown in Table 4.34. It is found that in all the states of this zone, the subsidy has increased except in Madhya Pradesh during pre as well as post-liberalisation periods. In Gujarat, subsidy has increased from Rs.24.82 crores in 1980-81 to Rs.2,562.58 crores in 2000-01 and further increased to Rs.3,551.51 crores in 2006-07. This has risen up from Rs.3.92 crores in 1980-81 to Rs.204.19 crores in 1990-91 and further rose to Rs. 1,311.24 crores in 2006-07 in Madhya Pradesh.

**Table 4.34: State-wise Distribution of Irrigation Subsidy in West Zone in India during 1980-81 to 2006-07 (In Rs. Crores)**

| *Years/States* | *West Zone* | | | | | |
|---|---|---|---|---|---|---|
| | *1980-81* | *1985-86* | *1990-91* | *1996-97* | *2000-01* | *2006-07* |
| Gujarat | 24.82<br>(6.22) | 188.43<br>(11.30) | 398.69<br>(10.18) | 1,730.34<br>(16.63) | 2,562.58<br>(17.42) | 3,551.51<br>(10.30) |
| Madhya Pradesh | 3.92<br>(0.98) | 55.9<br>(3.35) | 204.19<br>(5.21) | 453.43<br>(4.36) | 394.73<br>(2.68) | 1,311.24<br>(3.80) |
| Chhattisgarh | – | – | – | – | 51.5<br>(0.35) | 355<br>(1.03) |
| Maharashtra | 67.83<br>(17.00) | 255.94<br>(15.35) | 839.88<br>(21.44) | 2,138.11<br>(20.55) | 2,412.25<br>(16.40) | 5,643.65<br>(16.37) |
| Rajasthan | 33.11 | 158.51<br>(9.51) | 203.92<br>(5.21) | 651.46<br>(6.26) | 799.52<br>(5.43) | 1,350.85<br>(3.92) |
| Goa | – | – | – | 25.5<br>(0.25) | 45.9<br>(0.31) | 125.83<br>(0.36) |

*Source*: Government of India, Pricing of Water in Public System, 2010, Combined Finance and Revenue Accounts of different states.

*Note*: (1) Irrigation subsidies in Rs. Crores of different states are calculated by difference between expenditure and revenue of medium, minor as well as major irrigations.

(2) Percentages are shown in parentheses.

Chhattisgarh has got Rs.51.5 crores and Rs.355 crores in 2000-01 and 2006-07 respectively, whereas Goa received Rs.45.9 crores and Rs.125.83 crores in 2000-01 and 2006-07 respectively. In Maharashtra, irrigation subsidy has risen up from Rs.67.83 crores in 1980-81 to Rs.5643.65 crores in 2006-07. This type of subsidy has increased from Rs.33.11 crores in 1980-81 to Rs.651.46 crores in 1996-97 and further increased to Rs.1350.85 crores in 2006-07 in Rajasthan.

The percentage share analysis reveals that Maharashtra is ahead among all the other states by getting huge amount of subsidy during pre as well as post-liberalisation periods except in 2000-01. The percentage share of Goa has risen up during 1996-97 to 2006-07, whereas a lot of variation is seen in Gujarat, Madhya Pradesh and Chhattisgarh and Rajasthan.

In Gujarat, the percentage share at India level has increased from 6.22 in 1980-81 to 10.18 in 1990-91 and further increased to 17.42 in 2000-01. In Madhya Pradesh, it has increased from 0.98 in 1980-81 to 5.21 in 1990-91 and declined to 3.80 in 2006-07. Maharashtra has got 17.00 per cent, 21.44 per cent, 20.55 per cent and 16.37 per cent, on the other hand, Rajasthan has received 8.30 per cent, 5.21 per cent, 6.26 per cent and 3.92 per cent of irrigation subsidy in 1980-81, 1990-91, 1996-97 and 2006-07 respectively.

It is found that in all the states except in Madhya Pradesh, the irrigation subsidy has risen up in absolute terms during pre as well as post-liberalisation periods. As post-liberalisation period (2006-07) is compared to pre-liberalisation period (1990-91), in Gujarat this subsidy has increased approximately nine times more and in Maharashtra, in Rajasthan and in Madhya Pradesh increased by near about seven times. In 1990-91, Gujarat has got approximately two times more of irrigation subsidy as compared to Madhya Pradesh and Maharashtra has received more than four times as compared to Rajasthan.

**Table 4.35: State-wise Distribution of Irrigation Subsidy in North Zone in India during 1980-81 to 2006-07 (In Rs. Crores)**

| *Years/States* | *North Zone* | | | | | |
|---|---|---|---|---|---|---|
| | *1980-81* | *1985-86* | *1990-91* | *1996-97* | *2000-01* | *2006-07* |
| Haryana | 35.03<br>(8.78) | 70.67<br>(4.24) | 131.67<br>(3.36) | 373<br>(3.58) | 481.81<br>(3.28) | 903.7<br>(2.62) |
| Punjab | 9.55<br>(2.39) | 68.58<br>(4.11) | 192.54<br>(4.91) | 166.14<br>(1.60) | 467.39<br>(3.18) | 671.58<br>(1.95) |
| Uttar Pradesh | 40.37<br>(10.12) | 226.52<br>(13.59) | 637.51<br>(16.27) | 949.39<br>(9.12) | 1283.7<br>(8.73) | 2826<br>(8.20) |
| Jammu & Kashmir | 2.91<br>(0.73) | 15.95<br>(0.96) | 34.05<br>(0.87) | 20.79<br>(0.20) | 39.17<br>(0.27) | 72.29<br>(0.21) |
| Himachal Pradesh | 0.19<br>(0.05) | 4.48<br>(0.27) | 9.62<br>(0.25) | 5.77<br>(0.06) | 15.78<br>(0.11) | 44.58<br>(0.13) |
| Uttarakhand | – | – | – | – | 25.65<br>(0.17) | 237.42<br>(0.69) |

Source: Government of India, Pricing of Water in Public System, 2010, Combined Finance and Revenue Accounts of different states.

Note: (1) Irrigation subsidies in Rs. Crores of different states are calculated by difference between expenditure and revenue of medium, minor as well as major irrigations.

(2) Percentages are shown in parentheses.

The irrigation subsidy in north zone of India during 1980-81 to 2006-07 is shown in Table 4.35. The irrigation subsidy in Rs. Crores of different states is calculated by difference between expenditure and revenue of medium, minor as well as major irrigations. This table reveals that in all the states irrigation subsidy has increased except in Punjab, Jammu & Kashmir and Himachal Pradesh (in these states the irrigation has declined in 1996-97).

In Haryana, this has increased from Rs.35.03 crores in 1980-81 to Rs.481.81 crores in 2000-01 and further increased to Rs.903.7 crores in 2006-07. In Punjab, subsidy has risen up from Rs.9.55 crores in 1980-81 to Rs. 192.14 crores in 1990-91 and declined to Rs.166.14 crores in 1996-97 and again risen up to Rs.671.568 crores in 2006-07. Uttar Pradesh has got Rs.40.37 crores, Rs.637.51 crores and Rs.2,826 crores in 1980-81, 1990-91 and 2006-07 respectively.

In Jammu & Kashmir, this has increased from Rs.2.91 crores in 1980-81 to Rs.34.05 crores in 1990-91 and declined to Rs. 20.79 crores in 1996-97 and again increased to Rs.72.29 crores in 2006-07, whereas subsidy has increased from Rs.0.19 crores in 1980-81 to Rs.9.62 crores in 1990-91 and declined to Rs.5.77 crores in 1996-97 and again increased to Rs.44.58 crores in 2006-07 in Himachal Pradesh.

The percentage share of Haryana declined from 8.78 in 1980-81 to 2.62 in 2006-07. The percentage share of Punjab has declined from 2.39 in 1980-81 to 1.60 in 1996-97 and increased to 3.18 in 2000-01, on the other hand in Uttar Pradesh, it has risen up from 10.12 per cent in 1980-81 to 16.27 per cent in 1990-91 and declined to 9.12 per cent in 1996-97 and further declined to 8.20 per cent in 2006-07. In Jammu & Kashmir, the percentage share has increased from 0.73 in 1980-81 to 0.87 in 1990-91 and declined to 0.21 in 2006-07. It increased from 0.05 per cent in 1980-81 to 0.25 per cent in 1990-91 and declined to 0.13 per cent in 2006-07 in Himachal Pradesh.

From the above table, it is concluded that in all the states of north zone, irrigation subsidy has increased except in Punjab, Jammu & Kashmir and Himachal Pradesh during pre as well as post-liberalisation periods. As the year 2006-07 is compared to the year 1990-91, it is observed that in Haryana, this has increased near about seven times, in Uttar Pradesh as well as in Himachal Pradesh more than four times, in Jammu Kashmir more than two times and in Punjab more than three times. As compared to Haryana in 1990-91, Uttar Pradesh has got 4.84 times, whereas in 2006-07, it has got 3.12 times more of irrigation subsidy and as compared to Himachal Pradesh, Jammu & Kashmir has got more than three times and 1.62 times more of irrigation subsidy in 1990-91 and 2006-07 respectively.

The irrigation subsidy in east zone of India during 1980-81 to 2006-07 is shown in Table 4.36. It is observed that in all the state of this zone the irrigation subsidy has increased throughout the study period.

In Bihar, this has increased from Rs.6.09 crores in 1980-81 to Rs.276.84 in 1996-97 and further increased to Rs.624.95 in 2006-07. In Odisha, irrigation subsidy has risen up from Rs.9.52 crores in 1980-81 to Rs.691.9 crores in 2006-07, whereas this has increased from Rs.18.07 crores in 1980-81 to Rs.219.96 crores in 1996-97 and further increased to Rs.297.4 crores in 2006-07 in West Bengal.

The percentage share has increased from 1.53 in 1980-81 to 4.64 in 1990-91 and declined to 1.821 in 2006-07 in Bihar. The percentage share has declined from 2.39 in 1980-81 to 1.46 in 1990-91 and increased to 2.01 in 2006-07 in Odisha, whereas in West Bengal, it has increased from 4.53 per cent in 1980-81 to 5.00 per cent in 1985-

86 and to 0.86 per cent in 2006-07. It has increased from 0.37 per cent in 2000-01 to 0.72 per cent in 2006-07 in Jharkhand.

From the above table, it is concluded that in all the states (except in West Bengal) of east zone, the irrigation subsidy has increased in absolute terms throughout the study period. As post-liberalisation period (2006-07) is compared to pre-liberalisation period (1990-91), it is observed that in Odisha, this has risen up the maximum *i.e.* 12.8 times, Bihar more than three time and West Bengal more than two times. In 1990-91, Bihar has got 3.2 times more and West Bengal more than two times than that of Odisha, whereas in 2006-07. Odisha has received more than two times and 1.1 times more of irrigation subsidy as compared to Bihar and West Bengal respectively.

**Table 4.36: State-wise Distribution of Irrigation Subsidy in East Zone in India during 1980-81 to 2006-07 (In Rs. Crores)**

| *Years/States* | *East Zone* | | | | | |
|---|---|---|---|---|---|---|
| | *1980-81* | *1985-86* | *1990-91* | *1996-97* | *2000-01* | *2006-07* |
| Bihar | 6.09 (1.53) | 78.19 (4.69) | 181.93 (4.64) | 276.84 (2.66) | 484.29 (3.29) | 624.95 (1.81) |
| Jharkhand | – | – | – | – | 53.77 (0.37) | 246.95 (0.72) |
| Odisha | 9.52 (2.39) | 23.85 (1.43) | 57.01 (1.46) | 390.14 (3.75) | 469.71 (3.19) | 691.9 (2.01) |
| West Bengal | 18.07 (4.53) | 83.36 (5.00) | 145.32 (3.71) | 219.96 (2.11) | 298.51 (2.03) | 297.4 (0.86) |

*Source*: Government of India, Pricing of Water in Public System, 2010, Combined Finance and Revenue Accounts of different states.

*Note*: (1) Irrigation subsidies in Rs. Crores of different states are calculated by difference between expenditure and revenue of medium, minor as well as major irrigations.

(2) Percentages are shown in parentheses.

The irrigation subsidy of north-east zone of India during 1980-81 to 2006-07 is shown in Table 4.37. A lot of variation is seen in all the states of north-east zone throughout the study period. In Assam, this has increased from Rs.7.69 crores in 1980-81 to Rs.68.86 in 2000-01 and declined to Rs.59.14 crores in 2006-07. This has risen up from Rs.0.16 crores in 1980-81 to Rs.11.29 crores in 1990-91 and declined to Rs.4.13 crores in 1996-97 and again increased to Rs.10.5 crores in 2006-07 in Tripura.

In Manipur, this has increased from Rs.0.81 crores in 1980-81 to Rs.40.17 crores in 1996-97 and 27.41 crores in 2000-01 and again increased to Rs. 231.53 crores in 2006-07. Meghalaya has got Rs.1.61 crores, Rs.2.49 crores and Arunachal Pradesh received Rs.0.48 crores, 0.34 crores in 1996-97 and 2000-01 respectively.

The percentage share of Assam at India level declined from 1.93 in 1980-81 to 1.12 in 1990-91 and further to 0.17 in 2006-07. In Tripura, it has increased from 0.04 in 1980-81 to 0.29 in 1990-91 and declined to 0.03 in 2006-07, whereas it has declined from 0.20 in 1980-81 to 0.19 in 1985-86 and increased to 0.67 in 2006-07 in Manipur.

It is found that the percentage share of Meghalaya (0.02) remains constant in both the years 1996-97 and in 2000-01, whereas the percentage share has increased from 0.05 in 1980-81 to 0.07 in 1985-86 and declined to 0.0001 in 2000-01 and further declined to 0.00003 in 2006-07 in Mizoram.

Above table shows that in absolute terms, irrigation subsidy has declined in all the states of north-east zone, whereas a lot of variation is seen during the study period. As 2006-07 is compared to 1990-91, it is observed that in Manipur, this has increased more than fourteen times and in Assam 1.3 times. In 1990-91, Assam has got near about four times more of irrigation subsidy and near about six times, whereas Manipur 1.45 times and 22.05 times than that of Tripura in 1990-91 and 2006-07 respectively.

**Table 4.37: State-wise Distribution of Irrigation Subsidy in North-East Zone in India during 1980-81 to 2006-07 (In Rs. Crores)**

| *Years/States* | *North-East Zone* | | | | | |
|---|---|---|---|---|---|---|
| | *1980-81* | *1985-86* | *1990-91* | *1996-97* | *2000-01* | *2006-07* |
| Assam | 7.69 (1.93) | 21.65 (1.30) | 43.83 (1.12) | 28.01 (0.27) | 68.86 (0.47) | 59.14 (0.17) |
| Tripura | 0.16 (0.04) | 3.34 (0.20) | 11.29 (0.29) | 4.13 (0.04) | 7.18 (0.05) | 10.5 (0.03) |
| Manipur | 0.81 (0.20) | 3.09 (0.19) | 16.41 (0.42) | 40.17 (0.39) | 27.41 (0.19) | 231.53 (0.67) |
| Meghalaya | – | – | – | 1.61 (0.02) | 2.49 (0.02) | – |
| Mizoram | – | 0.81 (0.05) | 2.77 (0.07) | – | 0.01 (0.0001) | – |
| Arunachal Pradesh | – | – | – | 0.48 (0.005) | 0.34 (0.002) | 0.56 (0.002) |

*Source*: Government of India, Pricing of Water in Public System, 2010, Combined Finance and Revenue Accounts of different states.

*Note*: (1) Irrigation subsidies in Rs. Crores of different states are calculated by difference between expenditure and revenue of medium, minor as well as major irrigations.

(2) Percentages are shown in parentheses.

The per hectare irrigation subsidy of five zones of India 1980-81 to 2006-07 is shown in Table 4.38. The irrigation subsidy also increased in all the zones except in north-east zone (in this zone, this subsidy has declined in 2000-01) of India.

This table reveals that in India, this subsidy has increased during 1980-81 to 2006-07. This has gone up from Rs.430.27 crores in 1980-81 to Rs. 6,13,552.27 crores in 2006-07. In south this has increased from Rs.155.94 crores in 1980-81 to Rs.4,950.01 crores in 2000-01 and further increased to Rs.12,851.27 crores in 2006-07, whereas this subsidy risen up from Rs.44.12 crores in 1980-81 to Rs.12,730.72 crores in 2006-07 in west zone.

In north zone, this has gone up from Rs.126.57 crores in 1980-81 to Rs.21,325.23 crores in 2006-07, on the other hand in east zone irrigation subsidy has increased from Rs.40.06 in 1980-81 to Rs.3,896.11 in 2006-07. This has increased from Rs.63.57 crores in 1980-81 to Rs.2,225.34 crores in 1996-97 and declined to Rs.1,741.93 in crores 2000-01 and again increased to Rs.1,0548.93 crores in 2006-07 in north-east zone.

The percentage-wise analysis shows a lot of variation in all the zones. The percentage share of south zone has declined from 36.24 in 1980-81 to 25.98 in 1996-97 and further declined 20.95 in 2006-07. West zone has got 10.25 per cent, 12.17 per cent, 28.79 per cent and 20.75 per cent in 1980-81, 1990-91, 1996-97 and 2006-07 respectively. It is found that the percentage share of north zone has declined during 1980-81 to 2000-01 and increased in 2006-07. This zone has received 28.42 per cent, 24.37 per cent and 14.00 per cent in 1980-81, 1990-91, 1996-97 and 2006-07 respectively. It is observed that it has increased from 9.31 per cent in 1980-81 to 8.49 per cent in 1996-97 and declined to 6.35 per cent in 2006-07 in east zone.

**Table 4.38: Zone-wise Distribution of Irrigation Subsidy in India during 1980-81 to 2006-07 (In Rs./Hectare)**

| *Years/Zones* | *1980-81* | *1985-86* | *1990-91* | *1996-97* | *2000-01* | *2006-07* |
|---|---|---|---|---|---|---|
| South | 155.94<br>(36.24) | 443.43<br>(23.51) | 944.47<br>(19.98) | 3,033.41<br>(25.98) | 4,950.01<br>(30.26) | 12,851.27<br>(20.95) |
| West | 44.12<br>(10.25) | 306.28<br>(16.24) | 575.53<br>(12.17) | 3,361.04<br>(28.79) | 5,814.72<br>(35.55) | 12,730.72<br>(20.75) |
| North | 126.57<br>(29.42) | 513.15<br>(27.20) | 1,152.05<br>(24.37) | 2,064.44<br>(17.68) | 2,443.41<br>(14.00) | 21,325.23<br>(34.76) |
| East | 40.06<br>(9.31) | 204.47<br>(10.84) | 400.70<br>(8.48) | 990.65<br>(8.49) | 1,405.63<br>(8.59) | 3,896.11<br>(6.35) |
| North-East | 63.57<br>(14.77) | 418.96<br>(22.21) | 1,655.13<br>(35.01) | 2,225.34<br>(19.06) | 1,741.93<br>(10.65) | 10,548.93<br>(17.19) |
| India | 430.27<br>(100) | 1,886.30<br>(100) | 4,727.89<br>(100) | 11,674.89<br>(100) | 16,355.69<br>(100) | 61,352.27<br>(100) |

*Source*: (1) Government of India, Pricing of Water in Public System, 2010, Combined Finance and Revenue Accounts of different states.

(2) Government of Punjab, Statistical Abstract, various years.

*Note*: (1) Irrigation subsidies per hectare of zones are calculated by adding the irrigation subsidies on zone basis

(2) Percentages are shown in parentheses.

Above table shows that at national level as well as zone level (except in north-east zone), irrigation subsidy per hectare has increased in absolute term during pre as well as post-liberalisation periods. As post-liberalisation period (2006-07) as compared to pre-liberalisation period (1990-91), at country level this has increased near about twenty times, whereas zone-wise analysis reveals that in west zone, this has increased the maximum times *i.e.* twenty two times, in north zone 18.5 times, in south zone more than thirteen times, in east zone more than nine times and in north-east zone more than six times. As compared to east zone, the south zone has got 2.36 times more of irrigation subsidy (in 1990-91) and 3.3 times (in 2006-07), whereas north zone two times more (in 1990-91) and near about two times (in 2006-07) as compared to west zone.

The irrigation subsidy per hectare in five states of south zone during 1980-81 to 2006-07 is shown in Table 4.39. The irrigation subsidy per hectare of states is calculated by dividing the irrigation subsidy in Rs. Crores with gross cropped area of the concerned state. This table reveals that during 1980-81 to 1990-91 and 2000-01 to 2006-07, Andhra Pradesh has ranked first, whereas Karnataka has got first rank in 1996-97.

**Table 4.39: State-wise Distribution of Irrigation Subsidy in South Zone in India during 1980-81 to 2006-07 (In Rs./Hectare)**

| *Years/States* | *South Zone* | | | | | |
|---|---|---|---|---|---|---|
| | *1980-81* | *1985-86* | *1990-91* | *1996-97* | *2000-01* | *2006-07* |
| Andhra Pradesh | 48.29 | 155.70 | 281.06 | 973.22 | 1,623.88 | 8,240.72 |
| | (11.22) | (8.25) | (5.94) | (8.34) | (9.93) | (13.43) |
| Karnataka | 44.93 | 126.26 | 184.09 | 1,012.95 | 1,409.78 | 3,049.08 |
| | (10.44) | (6.69) | (3.89) | (8.68) | (8.62) | (4.97) |
| Kerala | 24.25 | 79.17 | 272.28 | 574.47 | 595.17 | 762.06 |
| | (5.64) | (4.20) | (5.76) | (4.92) | (3.64) | (1.24) |
| Tamil Nadu | 38.48 | 82.30 | 207.04 | 311.01 | 960.57 | 799.40 |
| | (8.94) | (4.36) | (4.38) | (2.66) | (5.87) | (1.30) |
| Pondicherry | – | – | – | 161.76 | 360.61 | – |
| | | | | (1.39) | (2.20) | |

*Source*: (1) Government of India, Pricing of Water in Public System, 2010, Combined Finance and Revenue Accounts of different states.

(2) Government of Punjab, Statistical Abstract, various years.

*Note*: (1) Irrigation subsidies per hectare of states are calculated by dividing the irrigation subsidies in Rs. Crores with gross cropped area of the concerned state.

(2) Percentages are shown in parentheses.

In Andhra Pradesh, this has increased from Rs.48.29 in 1980-81 to Rs.1,623.88 in 2000-01 and further increased to Rs.8,240.72 in 2006-07. In Karnataka, this subsidy has risen up from Rs.44.93 in 1980-81 to Rs.3,049.08 in 2006-07, whereas in Kerala, this subsidy has gone up from Rs.24.25 in 1980-81 to Rs.762.06 in 2006-07. Tamil Nadu has got Rs.38.48, Rs.207.04, Rs.311.01, and Rs.799.40 in 1980-81, 1990-91, 1996-97 and 2006-07 respectively.

A lot of variation is seen (zone-wise) in proportionate analysis of irrigation subsidy at India level during 1980-81 to 2006-07. The percentage share of Andhra Pradesh has declined from 11.22 in 1980-81 to 5.94 in 1990-91 and increased to 13.43 in 2006-07, whereas in Karnataka, it has declined from 10.44 per cent in 1980-81 to 3.89 per cent in 1990-91 and increased to 4.97 per cent in 2006-07. The percentage share of irrigation subsidy has been percentage share has declined from 5.64 in 1980-81 to 4.92 in 1996-97 and further declined to 1.24 in 2006-07 in Kerala.

In Tamil Nadu, it has also declined from 8.54 per cent in 1980-81 to 4.36 per cent in 1985-86 and increased to 4.38 per cent in 1990-91 and again declined to 1.30 per cent in 2006-07, whereas Pondicherry has received 1.39 per cent and 2.20 per cent of irrigation subsidy in 1996-97 to 2000-01 respectively.

It is observed that irrigation subsidy per hectare in all the states except in Tamil Nadu of south zone has increased in absolute terms throughout the study period. As the year 2006-07 is compared to the year 1990-91, in Andhra Pradesh, this has risen up by twenty nine times, in Karnataka sixteen times, in Tamil Nadu approximately four times and in Kerala near about three times. In 1990-91, Andhra Pradesh as well as Kerala has got 1.53 times and 1.32 times more of irrigation subsidy than that of Karnataka and Tamil Nadu respectively, whereas in 2006-07, Andhra Pradesh has received near about three times more and Tamil Nadu near about two times as compared to Karnataka and Kerala respectively.

The subsidy of irrigation in west zone of India during 1980-81 to 2006-07 is shown in Table 4.40. This table reveals that in all the states except in Rajasthan, the irrigation subsidy has increased throughout the study period.

**Table 4.40: State-wise Distribution of Irrigation Subsidy in West Zone in India during 1980-81 to 2006-07 (In Rs./Hectare)**

| *Years/States* | *West Zone* | | | | | |
|---|---|---|---|---|---|---|
| | *1980-81* | *1985-86* | *1990-91* | *1996-97* | *2000-01* | *2006-07* |
| Gujarat | 23.21<br>(5.39) | 194.60<br>(10.32) | 384.80<br>(8.14) | 1,572.89<br>(13.47) | 2,397.17<br>(14.66) | 2,772.23<br>(4.520) |
| Madhya Pradesh | 1.83<br>(0.43) | 24.29<br>(1.29) | 85.51<br>(1.81) | 185.44<br>(1.59) | 220.89<br>(1.35) | 1,054.22<br>(1.72) |
| Chhattisgarh | – | – | – | – | 96.68<br>(0.59) | 1,216.59<br>(1.98) |
| Rajasthan | 19.08<br>(4.44) | 87.40<br>(4.63) | 105.22<br>(2.23) | 314.82<br>(2.70) | 415.77<br>(2.54) | 371.98<br>(0.61) |
| Goa | – | – | – | 1,287.88<br>(11.03) | 2,684.21<br>(16.41) | 7,315.70<br>(11.92) |

*Source:* (1) Government of India, Pricing of Water in Public System, 2010, Combined Finance and Revenue Accounts of different states.

(2) Government of Punjab, Statistical Abstract, various years.

*Note:* (1) Irrigation subsidies per hectare of states are calculated by dividing the irrigation subsidies in Rs. Crores with gross cropped area of the concerned state.

(2) Percentages are shown in parentheses.

The subsidy has increased from Rs.23.21 in 1980-81 to Rs.2,397.17 in 2000-01 and further increased to Rs.2,772.23 in 2006-07. In Madhya Pradesh, subsidy has risen up from Rs.1.83 in 1980-81 to Rs.1,054.22 in 2006-07, on the other hand Chhattisgarh has got Rs.96.68 and Rs.1,216.59 in 2000-01 and 2006-07 respectively.

In Maharashtra, the irrigation subsidy has gone up from Rs.33.46 in 1980-81 to Rs.6,926.42 in 2006-07. In Rajasthan, this has increased from Rs.19.08 in 1980-81 to Rs.371.98 in 2006-07. In Goa, subsidy of irrigation has gone up by 108.42 per cent in 2000-01 and 172.55 per cent in 2006-07 as compared to 1996-97 and 2000-01 respectively.

In Gujarat, the percentage share has increased from 5.39 in 1980-81 to 8.14 in 1990-91 and declined to 4.52 in 2006-07. In Madhya Pradesh, the percentage share has risen up from 0.43 in 1980-81 to 1.59 in 1996-97 and further increased to 1.72 in 2006-07. Rajasthan has got 4.44 per cent, 2.23 per cent, 2.70 per cent and 0.61 per cent in 1980-81, 1990-91, 1996-97 and 2006-07 respectively, whereas the percentage share risen up from 11.03 in 1996-97 to 16.41 in 2000-01 and declined to 11.92 in 2006-07 in Goa.

It is found that in absolute terms, the irrigation subsidy per hectare has increased in all states of west zone during pre as well as post-liberalisation periods. As the year 2006-07 is compared to the year 1990-91, it is observed that in Madhya Pradesh, this has increased more than twelve times and in Gujarat seven times. It is found that in 1990-91, Gujarat has 4.5 four times, whereas in 2006-07, 2.63 times more of irrigation subsidy as compared to Madhya Pradesh. In 1996-97, Goa has received more than four times and in 2006-07, 19.66 times more than that of Rajasthan.

The irrigation subsidy per hectare in north zone of India during 1980-81 to 2006-07 is shown in Table 4.41. It is seen that Haryana is only state of this zone in which irrigation subsidy has increased during pre and post-liberalisation periods.

**Table 4.41: State-wise Distribution of Irrigation Subsidy in North Zone India in during 1980-81 to 2006-07 (In Rs./Hectare)**

| *Years/States* | *North Zone* | | | | | |
|---|---|---|---|---|---|---|
| | *1980-81* | *1985-86* | *1990-91* | *1996-97* | *2000-01* | *2006-07* |
| Haryana | 64.13<br>(14.91) | 126.17<br>(6.69) | 222.45<br>(4.71) | 614.09<br>(5.26) | 787.91<br>(4.82) | 1,413.36<br>(2.30) |
| Punjab | 14.12<br>(3.28) | 95.81<br>(5.08) | 256.65<br>(5.43) | 826.69<br>(7.08) | 663.29<br>(4.06) | 167,76.45<br>(27.34) |
| Uttar Pradesh | 16.43<br>(3.82) | 90.32<br>(4.79) | 250.20<br>(5.29) | 363.35<br>(3.11) | 474.44<br>(2.90) | 109.53<br>(0.18) |
| Jammu & Kashmir | 29.88<br>(6.94) | 154.85<br>(8.21) | 319.42<br>(6.76) | 193.04<br>(1.65) | 351.30<br>(2.15) | 642.01<br>(1.05) |
| Himachal Pradesh | 2.01<br>(0.47) | 46.00<br>(2.44) | 103.33<br>(2.19) | 60.93<br>(0.52) | 166.46<br>(1.02) | 470.75<br>(0.77) |
| Uttarakhand | – | – | – | – | – | 1,913.13<br>(3.12) |

*Source*: (1) Government of India, Pricing of Water in Public System, 2010, Combined Finance and Revenue Accounts of different states.

(2) Government of Punjab, Statistical Abstract, various years.

*Note*: (1) Irrigation subsidies per hectare of states are calculated by dividing the irrigation subsidies in Rs. Crores with gross cropped area of the concerned state.

(2) Percentages are shown in parentheses.

In Haryana state, this has increased from Rs.64.13 in 1980-81 to Rs.614.09 in 1996-97 and further increased to Rs.1,413.36 in 2006-07. This increased from Rs.14.12 in 1980-81 to Rs.826.69 in 1996-97 and declined to Rs.663.29 in 2000-01 and again increased to Rs.16,776.45 in 2006-07 in Punjab.

In Uttar Pradesh, irrigation subsidy has gone up from Rs.16.43 in 1980-81 to Rs.363.35 in 1996-97 and further gone up to Rs.474.44 in 2000-01 and declined to Rs.109.53 in 2006-07. In Jammu & Kashmir, irrigation subsidy has risen up from Rs.29.88 in 1980-81 to Rs.319.42 in 1990-91 and declined to Rs.193.04 in 1996-97 and again risen up to Rs.642.01 in 2006-07. This has increased from Rs.2.01 in 1980-81 to Rs.103.33 in 1990-91 and declined to Rs.60.93 in 1996-97 and again increased to Rs.470.75 in 2006-07 in Himachal Pradesh.

The percentage share in Haryana has declined from 14.91 in 1980-81 to 4.71 in 1990-91 and further declined to 2.30 in 2006-07, whereas in Punjab, it has increased from 3.28 in 1980-81 to 5.43 in 1990-91 and further to 27.34 in 2006-07. The percentage share of irrigation subsidy per hectare increased from 3.82 in 1980-81 to 5.29 in 1990-91 and declined to 0.18 in 2006-07 in Uttar Pradesh.

From the above analysis, it is found that in all the states irrigation subsidy per hectare has declined except in Haryana throughout the study period. As post-liberalisation period (2006-07) is compared to pre-liberalisation period (1990-91), it is observed that in Punjab, this has increased the maximum *i.e.* more than sixty five times, in Haryana more than six times and in Himachal Pradesh more than four times. In 1990-91, Uttar Pradesh has got 1.1 times more of irrigation subsidy than that of Haryana, whereas in 2006-07, this has increased by approximately thirteen times as compared to Uttar Pradesh. On the other hand, Jammu & Kashmir has got three times more and 1.36 times in 1990-91 and 2006-07 as compared to Himachal Pradesh.

The irrigation subsidy of east zone during 1980-81 to 2006-07 is shown in Table 4.42. Irrigation subsidy per hectare of states is calculated by dividing the irrigation subsidy in Rs. Crores with gross cropped area of the concerned state. This reveals that in four states of the same zone, this subsidy has increased during pre as well as post-liberalisation periods.

In Bihar, this has increased from Rs.5.46 in 1980-81 to Rs.824.25 in 2006-07, whereas in Odisha this has risen up from Rs.10.88 in 1980-81 to Rs.596.23 in 2000-01 and further rose to Rs.1,620.37 in 2006-07. West Bengal has got Rs.23.71, Rs.167.77, Rs.242.81 and Rs.517.13 of irrigation subsidy in 1980-81, 1990-91, 1996-97and 2006-07 respectively.

The percentage share of Bihar has increased from 1.27 in 1980-81 to 2.34 in 1996-97 and declined to 1.34 in 2006-07. In Odisha, the percentage share declined from 2.53 in 1980-81 to 1.26 in 1990-91 and increased to 4.07 in 1996-97 and again declined to 2.64 in 2006-07, on the other hand, the percentage share of irrigation subsidy has declined from 5.51 in 1980-81 to 0.84 in 2006-07 in West Bengal.

**Table 4.42: State-wise Distribution of Irrigation Subsidy in East Zone in India during 1980-81 to 2006-07 (In Rs./Hectare)**

| *Years/States* | *East Zone* | | | | | |
|---|---|---|---|---|---|---|
| | *1980-81* | *1985-86* | *1990-91* | *1996-97* | *2000-01* | *2006-07* |
| Bihar | 5.46<br>(1.27) | 74.35<br>(3.94) | 173.51<br>(3.67) | 272.99<br>(2.34) | 481.98<br>(2.95) | 824.25<br>(1.340 |
| Jharkhand | – | – | – | – | – | 934.35<br>(1.52) |
| Odisha | 10.88<br>(2.53) | 25.76<br>(1.37) | 59.42<br>(1.26) | 474.85<br>(4.07) | 596.23<br>(3.65) | 1,620.37<br>(2.64) |
| West Bengal | 23.71<br>(5.51) | 104.37<br>(5.53) | 167.77<br>(3.55) | 242.81<br>(2.08) | 327.42<br>(2.00) | 517.13<br>(0.84) |

*Source*: (1) Government of India, Pricing of Water in Public System, 2010, Combined Finance and Revenue Accounts of different states.

(2) Government of Punjab, Statistical Abstract, various years.

*Note*: (1) Irrigation subsidies per hectare of states are calculated by dividing the irrigation subsidies in Rs. Crores with gross cropped area of the concerned state.

(2) Percentages are shown in parentheses.

Above table reveals that during pre as well as post-liberalisation periods, irrigation subsidy per hectare has been increased in all the states of east zone in absolute terms, whereas a lot of variation is seen in percentage-wise analysis. In Odisha, this has increased the maximum *i.e.*, more than twenty seven times and in Bihar near about five and in West Bengal more than three times. In 1990-91, Bihar has received approximately three times than that of Odisha, whereas in 2006-07, Odisha has got 1.96 times more of irrigation subsidy as compared to Bihar.

The irrigation subsidy of north-east zone during 1980-81 to 2006-07 is shown in Table 4.43. Irrigation subsidy per hectare of states is calculated by dividing the irrigation subsidy in Rs. Crores with gross cropped area of the concerned state. Variations are seen in all the states except in Meghalaya during pre as well as post-liberalisation periods.

In Assam, this has increased from Rs.22.32 in 1980-81 to Rs.115.43 in 1990-91 and declined to Rs.70.36 in 1996-97 and again increased to Rs.166.45 in 2006-07. This has risen up from Rs.4.27 in 1980-81 to Rs.253.71 in 1990-91 and declined to Rs.167.76 in 2000-01 and again rose to Rs.0.50 in 2006-07 in Tripura.

In Manipur, irrigation subsidy has increased from Rs.36.99 in 1980-81 to Rs.911.67 in 1990-91 and further increased to Rs.10,022.94 in 2006-07. In Meghalaya, this subsidy has increased from Rs.65.98 in 1996-97 to Rs.79.30 in 2000-01, whereas this has declined from Rs.20.00 in 1996-97 to Rs.12.93 in 2000-01 and increased to Rs.20.36 in 2006-07 in Arunachal Pradesh.

The percentage share-wise analysis shows that in Assam, the percentage share declined from 5.19 in 1980-81 to 0.27 in 2006-07. The percentage share of Tripura increased from 0.99 in 1980-81 to 4.19 in 1985-86 and further increased to 5.37 in

1990-91 and declined to 0.77 in 1996-97 and further declined to 0.55 in 2006-07, whereas Manipur has got 8.60 per cent, 19.28 per cent, 16.95 per cent and 16.34 per cent in 1980-81, 1990-91, 1996-97 and 2006-07 respectively. It is observed that in Arunachal Pradesh as well as in Mizoram the percentage share has declined during post-liberalisation period.

A lot of variation is observed in absolute terms as well as in percentage-wise analysis of irrigation subsidy per hectare in all the states of north-east zone during the study period. As post-liberalisation period (2006-07) is compared to pre-liberalisation period (1990-91), in Manipur, this has risen up in the maximum 10.99 times, in Assam as well as in Tripura 1.34 times only. Manipur has got approximately eight times more of irrigation subsidy, whereas more than sixty times in 1990-91 and 2006-07 respectively as compared to Assam.

**Table 4.43: State-wise Distribution of Irrigation Subsidy in North-East Zone in India during 1980-81 to 2006-07 (In Rs./Hectare)**

| *Years/States* | *North-East Zone* | | | | | |
|---|---|---|---|---|---|---|
| | *1980-81* | *1985-86* | *1990-91* | *1996-97* | *2000-01* | *2006-07* |
| Assam | 22.32<br>(5.19) | 57.06<br>(3.03) | 115.43<br>(2.44) | 70.36<br>(0.60) | 169.40<br>(1.04) | 166.45<br>(0.27) |
| Tripura | 4.27<br>(0.99) | 78.96<br>(4.19) | 253.71<br>(5.37) | 90.17<br>(0.77) | 167.76<br>(1.03) | 338.71<br>(0.55) |
| Manipur | 36.99<br>(8.60) | 168.85<br>(8.95) | 911.67<br>(19.28) | 1,978.82<br>(16.95) | 1,311.48<br>(8.02) | 10,022.94<br>(16.34) |
| Meghalaya | – | – | – | 65.98<br>(0.57) | 79.30<br>(0.48) | – |
| Mizoram | – | 114.08<br>(6.05) | 374.32<br>(7.92) | – | 1.06<br>(0.01) | 0.47<br>(0.001) |
| Arunachal Pradesh | – | – | – | 20.00<br>(0.17) | 12.93<br>(0.08) | 20.36<br>(0.03) |

*Source*: (1) Government of India, Pricing of Water in Public System, 2010, Combined Finance and Revenue Accounts of different states.

(2) Government of Punjab, Statistical Abstract, various years.

*Note*: (1) Irrigation subsidies per hectare of states are calculated by dividing the irrigation subsidies in Rs. Crores with gross cropped area of the concerned state.

(2) Percentages are shown in parentheses.

It is observed from the analysis that at national level, the increasing rate of total subsidies (fertilizers, electricity and irrigation) is higher than gross cropped area (GCA) during pre, first as well as second phase of liberalization periods. In 1985-86, the total subsidies have increased by 290.39 per cent and gross cropped area by 2.4 per cent as compared from 1980-81, in 1996-97, total subsidies have increased by 174.78 per cent and GCA by 1.73 per cent as from 1990-91, whereas total subsidies have increased by 109.53 per cent in 2008-09 and GCA declined by 5.84 per cent in 2006-07 as compared to 2000-01 at the national level.

In total subsidies during pre-liberalisation period, the percentage share of fertilizers subsidies is maximum (38.41 in 1980-81 and 37.63 in 1985-86), whereas during 1990-91 to 2000-01, the percentage share of electricity subsidies is maximum and again in 2008-09 fertilizers subsidies has got major percentage share 87.26 per cent in total subsidies.

There is a lot of variation to find out the relationship between gross cropped area (GCA) and in total subsidies in zones as well as in states throughout the study period. As zone level, it is observed that there is a negative relationship between GCA and total subsidies, in west zone and in north zone (in 2006-07) and in east zone (1996-97 and in 2006-07) and at state level in Andhra Pradesh (in 1985-86 and in 2006-07), Karnataka (in 2000-01), Kerala, Punjab, West Bengal, Madhya Pradesh and Assam (in 2006-07), Tamil Nadu (during 1990-91 to 2000-01), Gujarat and Mizoram (in 1985-86), Maharashtra (in 1996-97), Rajasthan and Tripura (in 2000-01), Bihar (during 1985-86 to 2006-07), Odisha (during 1996-97 to 2006-07), Manipur (in 1985-86 and in 1990-91), the GCA has declined, whereas total subsidies have increased.

It is seen that there is a direct relationship in GCA and total subsidies *i.e.* GCA as well as total subsidies have increased at zone level in west and in north (during 1980-81 to 2000-01), in south zone (during 1980-81 to 1996-97) and in north-east zone (during 1980-81 to 2006-07) and at state level, in Andhra Pradesh, Gujarat and West Bengal (in 1990-91), Tamil Nadu (in 1985-86), Karnataka, Haryana (in 1985-86 and in 2006-07), Madhya Pradesh, Rajasthan (in 1996-97), Maharashtra (in 2000-01), Uttar Pradesh, Jammu & Kashmir and in Odisha (in 1985-86).

During pre-liberalisation period (in 1990-91), it is observed that Uttar Pradesh is leading among all the other states of India by receiving maximum percentage share of GCA *i.e.* 13.74 followed by Madhya Pradesh (12.88 per cent), Maharashtra (11.79 per cent), Rajasthan (10.45 per cent) and Andhra Pradesh and in post-liberalisation period (2006-07), Rajasthan is ahead among all the other states by getting 20.67 per cent of GCA followed by Uttar Pradesh (14.69 per cent), Andhra Pradesh and Gujarat (7.29 per cent), Karnataka and Madhya Pradesh (7.08 per cent) and Tamil Nadu (4.64 per cent). In total subsidies (in Rs. crores), Uttar Pradesh has ranked first, Maharashtra, Andhra Pradesh, Gujarat, Rajasthan and Punjab ranked second, third, fourth, fifth and sixth respectively in 1990-91, whereas Uttar Pradesh leading among all the states followed by Andhra Pradesh, Maharashtra, Punjab and Karnataka in 2008-09.

In fertilizers subsidies (in Rs. crores), Uttar Pradesh is ahead among all the by getting maximum percentage share 17.96 followed by Andhra Pradesh (12.98 per cent), Maharashtra (10.56 per cent), Punjab (9.79 per cent) and Madhya Pradesh (7.77 per cent) in 1990-91, on the other hand in 2008-09, Uttar Pradesh has got first position by receiving 16.19 per cent of fertilizers subsidies, followed by Andhra Pradesh (12.33 per cent), Maharashtra (10.30 per cent), Karnataka (7.35 per cent) and Punjab (7.10 per cent).

In 1990-91, Uttar Pradesh has ranked first by receiving maximum electricity subsidy (in Rs. crores) *i.e.* 15.21 per cent followed by Maharashtra (12.79 per cent), Gujarat (11.67 per cent), Punjab (10.81 per cent) and Andhra Pradesh (10.41 per cent), whereas Haryana, which has got 17.85 per cent of electricity subsidy, is leading the other states of India like Punjab (17.61 per cent), Tamil Nadu (12.40 per cent), Uttar Pradesh (10.37 per cent) and Karnataka (10.09 per cent) during post-liberalisation period (in 2008-09).

Maharashtra has got topmost position by getting 21.44 per cent of irrigation subsidy (in Rs. crores), Uttar Pradesh (16.27 per cent), Gujarat (10.18 per cent), Andhra Pradesh (9.46 per cent) and Karnataka (5.53 per cent) has got second, third, fourth and fifth position in pre-liberalisation period (in 1990-91). Whereas in post-liberalisation period (in 2006-07), it is observed that Andhra Pradesh (30.62 per cent), followed by Maharashtra (16.37 per cent), Karnataka (11.00 per cent), Gujarat (10.30 per cent) and Uttar Pradesh (8.20 per cent).

# Chapter 5

# Agricultural Subsidies and Productivity in India

Substantial additional growth in agricultural production needed to meet the basic necessities of large and growing population. It is also needed to generate agricultural surplus require for economic development with emphasis on employment equity. The bulk of growth is agricultural production will have to come from continuous increase in the productivity of land, yield based growth cannot sustain without removing soil fertility constraints and promote technological change. Among the various agriculture subsidies, fertilizer subsidy is the next largest food subsidy. Fertilizer subsidy is a development subsidy, which accelerate the fertilizer use and thus promote agricultural production. The central government removed the subsidy of fertilizer in the year 2003. There after agricultural production gradually decreased. The farmers were not able to purchased fertilizer on the higher price. In such a case farmers, fertilizer use for of their agricultural production gradually declined (Halmandage, 2010).

The overall rate of agricultural production is decreasing and production cost is increasing due to removal of agricultural subsidy. The removal of subsidy would affect the agricultural sector and economy. Subsidies are among the most powerful instrument for manipulating or balancing the growth rate of production and trade in various sectors for an equitable distribution of income for protection of weaker sections of the society. The support and procurement prices of major agricultural production are some of the important measures which are done to protect the interest of farmers and weaker sections of consumers. Substantial additional growth in agricultural production is needed to meet basic necessities for a larger growing population. It is also needed to generate agricultural surplus required for economic development with emphasis on employment equity. The agricultural production increased in initial period gradually after than the fertilizer subsidies were reduced.

The overall economy effected. The government policy of subsidy is very well for protection of the weaker sections and marginal farmers (Halmandage, 2010).

In this chapter, an attempt is made to analyse the relationship between total subsidies (including fertilizers, electricity and irrigation) and productivity. The whole analysis is divided into two sections. The section first deals with total subsidies and productivity at country as well as zone level during 1980-81 to 2006-07 and in section second, total subsidies and productivity at districts level of Punjab during 1980-81 to 2009-10 has been discussed.

## Section – I

Zone-wise distribution of total subsidies and productivity of crops in India during 1980-81 to 2006-07 are shown in Table 5.1. It is observed that at India level, the total subsidies have increased during the pre and post-liberalisation periods, whereas productivity increased except in 1996-97. At all India level, the total subsidies per hectare have increased from Rs. 1,362.59 in 1980-81 to Rs.14,134.59 (171.02 per cent) in 1990-91 and further increased to Rs.1,17,595.59 (85.38 per cent as compared to 2001). On the other hand, productivity in kgs. per hectare has increased from 4,67,491 in 1980-81 to 17,75,699 (44.63 per cent) in 1990-91 and declined to 14,56,191 (17.99 per cent) in 1996-97 and again increased to 19,53,815 (13.28 per cent) in 2006-07.

The zone wise comparison shows that in south zone, total subsidies have increased from Rs.439.59 in 1980-81 to Rs. 33,825.92 in 2006-07, whereas there is variation in change in the productivity *i.e.* it has increased from 76,335 kgs in 1980-81 to 4,96,733 kgs. in 1990-91 and declined to 4,06,587 kgs. in 1996-97 and again increased to 5,46,465 kgs. in 2006-07. The total subsidies have increased from Rs.219.34 in 1980-81 to Rs.29,069.85 in 2006-07, whereas the productivity has increased from 1,03,412 kgs. in 1980-81 to 3,76,109 kgs. in 1990-91 and declined to 3,29,723 kgs. in 1996-97 and again increased to 421954 kgs. in 2006-07 in West zone.

North zone has got Rs.489.09 as total subsides in 1980-81, which has increased to Rs.10,436.22 in 1996-97. This table shows that this zone has got subsides of Rs.14,200.76 and Rs. 31,536.06 in 2000-01 and 2006-07 respectively. The productivity of this zone has increased from 1,38,069 kgs. in 1980-81 to 3,65,489 kgs. in 1990-91 and declined to 3,17,125 kgs. in 1996-97 and again increased to 4,05,553 kgs. in 2006-07.

East zone received Rs.115.68 of subsides in 1980-81, increased to Rs.2,724.50 in 1996-97, whereas productivity of this zone has increased from 66,476 kgs. in 1980-81 to 2,57,507 kgs. in 1996-97 and declined to 2,36,814 kgs. in 2000-01 and again increased to 2,87,094 kgs. in 2006-07. The total subsidies have increased from Rs.98.89 in 1980-81 to Rs. 13,118.88 (284.05 per cent as compared to 2000-01) in 2006-07 and productivity has increased from 83,199 kgs. in 1980-81 to 2,92,749 kgs. in 2006-07 in north-east zone.

This table further reveals the percentage share of total subsidies in south zone at all India level, has declined from 32.26 in 1980-81 to 26.81 in 1990-91 and increased to 32.92 in 1996-97 and again declined to 28.76 in 2006-07, whereas the percentage share in productivity increased to 16.32 in 1980-81 to 31.18 in 1985-86 and declined

**Table 5.1: Zone-wise Distribution of Total Subsidies and Productivity during 1980-81 to 2006-07 (Subsidies (Subs.) in Rs./Hectare, Productivity (Prod.) in kgs./Hectare)**

| *Years* | *1980-81* | | *1985-86* | | *1990-91* | | *1996-97* | | *2000-01* | | *2006-07** | |
|---|---|---|---|---|---|---|---|---|---|---|---|---|
| *Subs. and Prod./Zones* | *Subs.* | *Prod.* | *Subs.* | *Prod.* | *Subs.* | *Prod.* | *Subs.* | *Prod.* | *Subs.* | *Prod.* | *Subs.* | *Prod.* |
| South | 439.59 (32.26) | 76,335 (16.33) | 1,478.37 (28.35) | 3,82,851 (31.18) | 3,788.86 (26.81) | 4,96,733 (27.97) | 12,856.34 (32.92) | 4,06,587 (27.92) | 23,282.67 (36.70) | 4,58,894 (26.61) | 33,825.92 (28.76) | 5,46,465 (27.97) |
| West | 219.34 (16.10) | 1,03,412 (22.12) | 949.44 (18.20) | 3,19,364 (26.01) | 2,733.66 (19.34) | 3,76,109 (21.18) | 10,147.02 (25.99) | 3,29,723 (22.64) | 18,004.45 (28.38) | 3,87,636 (22.48) | 29,069.85 (24.72) | 4,21,954 (21.60) |
| North | 489.09 (35.89) | 1,38,069 (29.53) | 1,676.93 (32.15) | 2,45,986 (20.04) | 4,213.34 (29.81) | 3,65,489 (20.58) | 10,436.22 (26.73) | 3,17,125 (21.78) | 14,200.76 (22.39) | 3,97,250 (23.03) | 31,536.06 (26.82) | 4,05,553 (20.76) |
| East | 115.68 (8.49) | 66,476 (14.22) | 544.61 (10.44) | 2,05,240 (16.72) | 1,308.94 (9.26) | 2,57,507 (14.50) | 2,724.50 (6.98) | 2,50,446 (17.20) | 4,530.41 (7.14) | 2,36,814 (13.73) | 10,044.87 (8.54) | 2,87,094 (14.69) |
| North-East | 98.89 (7.26) | 83,199 (17.80) | 566.01 (10.85) | 74,283 (6.05) | 2,089.79 (14.78) | 2,79,861 (15.76) | 2,884.02 (7.39) | 1,52,310 (10.46) | 3,415.94 (5.39) | 2,44,101 (14.15) | 13,118.88 (11.16) | 2,92,749 (14.98) |
| India | 1,362.59 (100) | 4,67,491 (100) | 5,215.36 (100) | 12,27,724 (100) | 14,134.59 (100) | 17,75,699 (100) | 39,048.10 (100) | 1,45,6191 (100) | 63,434.23 (100) | 1,72,4695 (100) | 1,17,595.59 (100) | 1,95,3815 (100) |

*Source*: (1) Government of Punjab, Statistical Abstract, various years; (2) Government of India, Fertilizers Association, New Delhi; (3) Government of India, State Electricity Boards (SEBs), Annual reports, various years; (4) Government of India, Water Price policy, 2010.

*Note*: (1) The total subsidies Rs. per hectare and Productivity in kgs. per hectare are calculated by adding the subsidies and productivity on zone basis; (2) *Subsidies of this year does not include irrigation subsidies; (3) Percentages are shown in parentheses.

to 26.61 in 2000-01 and again increased to 27.97 in 2006-07. The percentage share of west zone in total subsidies, has increased from 16.10 per cent in 1980-81 to 25.99 in 1996-97 and declined to 24.72 in 2006-07 and the percentage share in productivity increased from 22.12 in 1980-81 to 26.01 in 1985-86 and declined to 21.18 in 1990-91 and again increased to 21.60 in 2006-07. North zone has got first position by receiving 35.89 per cent share of total subsidies and 29.53 of productivity at all India level in 1980-81. Whereas it has got the same rank in case of subsidies in 1985-86 and 1990-91 and second in 1996-97, 2006-07 and third in 2000-01, on the other hand, it has got third position in 1985-86, 1990-91, 1996-97 and in 2006-07 in productivity.

East zone is that zone whose percentage share in productivity is more than its percentage share in subsidies during pre as well as post-liberalisation periods. A lot of variation is seen in east as well as north-east zones in subsidies and in productivity during pre and post-liberalisation periods (See the Table 5.1).

That table reveals that during pre as well as post-liberalisation periods, at country level as well as zone level, the total subsidies have increased in absolute terms, whereas at India level as well as in south, west, north, north-east zones, productivity has also increased except in 1996-97 and in east zone productivity has declined during 1996-97 to 2000-01. As compared to post-liberalisation period (2006-07) with pre-liberalisation period (1990-91), it is observed that in India, subsidies have increased 8.32 times, whereas productivity increased by only 1.1 times. While comparing the same time period, as zone level analysis shows that in west zone, subsidies have increased the maximum number of times *i.e.* 11.95 times, followed by south zone (8.93 times), east zone (7.67 times), north zone (7.49 times) and north-east zone (6.28 times), On the other hand productivity has increased maximum *i.e.* 1.90 times in south zone, followed by west zone (1.12 times), north zone (1.11 times), east zone (1.1 times) and north-east zone (1.05 times). In 1990-91, south zone has got near about three times of total subsidies and has near about two times of productivity, whereas in 2006-07, it has received 3.37 times of subsidies and has near about two times of productivity as compared to east zone.

The State-wise distribution of total subsidies and productivity of different states in south zone in India are shown in Table 5.2. This table reveals that in Andhra Pradesh, the total subsidies have increased from Rs.102.89 in 1980-81 to Rs.10,509.40 in 2006-07. The productivity of important crops in this state has increased from 16,791 kgs. in 1980-81 to 1,07,655 kgs. in 2006-07. Karnataka has received Rs.74.71, Rs.729.16 and Rs.4,446.72 in 1980-81, 1990-91 and 2006-07 respectively. In this state, the productivity has risen up from 22,760 kgs. in 1980-81 to 1,23,502 kgs. and declined to 1,04,829 kgs. in 2006-07. Kerala has received subsidies of Rs.56.25 in 1980-81, which have increased to Rs.1,599.75 in 2006-07.

The productivity of Kerala has increased from 7,885 kgs. in 1980-81 to 1,09,076 kgs. in 1985-96 and declined to 67,986 kgs. in 1990-91 again increased to 95,193 kgs. in 2000-01 and again declined to 90,560 kgs. in 2006-07. In 1980-81, the total subsides of Rs.205.73 are distributed in Tamil Nadu, which have increased to Rs.2,832.93 in 1996-97 and declined to Rs. 2,414.77 in 2006-07. On the other hand the productivity increased from 13,627 kgs. in 1980-81 to 1,29,466 kgs. in 1996-97 and declined to 1,31,527 kgs. in 2006-07.

**Table 5.2: State-wise Distribution of Total Subsidies and Productivity in South Zone in India during 1980-81 to 2006-07 (Subsidies (Subs.) in Rs./Hectare, Productivity (Prod.) in kgs./Hectare)**

| | South Zone | | | | | | | | | | | |
|---|---|---|---|---|---|---|---|---|---|---|---|---|
| *Years* | *1980-81* | | *1985-86* | | *1990-91* | | *1996-97* | | *2000-01* | | *2006-07** | |
| *Subs. and Prod./Zones* | *Subs.* | *Prod.* | *Subs.* | *Prod.* | *Subs.* | *Prod.* | *Subs.* | *Prod.* | *Subs.* | *Prod.* | *Subs.* | *Prod.* |
| Andhra Pradesh | 102.89 (7.55) | 16,791 (3.59) | 412.32 (7.91) | 75,200 (6.13) | 1,100.00 (7.78) | 83,104 (4.68) | 2,817.43 (7.22) | 8,3153 (5.71) | 5,660.91 (8.92) | 97,414 (5.65) | 10,509.40 (8.94) | 1,07,655 (5.51) |
| Karnataka | 74.71 (5.48) | 22,760 (4.87) | 323.89 (6.21) | 87,592 (7.13) | 729.16 (5.16) | 93,983 (5.29) | 2,371.84 (6.07) | 1,06,349 (7.30) | 3,855.76 (6.08) | 1,23,502 (7.16) | 4,446.72 (3.78) | 1,04,829 (5.37) |
| Kerala | 56.25 (4.13) | 7,885 (1.69) | 185.85 (3.56) | 1,09,076 (8.88) | 601.13 (4.25) | 67,986 (3.83) | 1,036.92 (2.66) | 82,474 (5.66) | 1,386.13 (2.19) | 95,193 (5.52) | 1,599.75 (1.36) | 90,560 (4.64) |
| Tamil Nadu | 205.73 (15.10) | 13,627 (2.91) | 556.30 (10.67) | 1,06,759 (8.70) | 1,358.58 (9.61) | 1,28,040 (7.21) | 2,832.93 (7.25) | 1,29,466 (8.89) | 6,061.02 (9.55) | 1,35,192 (7.84) | 2,414.77 (2.05) | 1,31,527 (6.73) |
| Pondicherry | – | 12,717 (2.72) | – | 2,344 (0.19) | – | 96,112 (5.41) | 3,496.26 (8.95) | 2,489 (0.17) | 6,243.98 (9.84) | 4,841 (0.28) | 14,345.50 (12.20) | 80,998 (4.15) |
| Andaman and Nicobar Islands | – | 2,555 (0.55) | – | 1,880 (0.15) | – | 27,508 (1.55) | 28.95 (0.07) | 2,656 (0.18) | 74.87 (0.12) | 2,752 (0.16) | 509.79 (0.43) | 30,896 (1.58) |

*Source*: (1) Government of Punjab, Statistical Abstract, various years; (2) Government of India, Fertilizers Association, New Delhi; (3) Government of India, State Electricity Boards (SEBs), Annual reports, various years; (4) Government of India, Water Price policy, 2010.

*Note*: (1) * Subsidies of this year does not include irrigation subsidies; (2) Percentages are shown in parentheses.

Andaman and Nicobar Islands has got 158.65 per cent more subsidies in 2000-01 and 580.90 per cent in 2006-07 as compared to the predecessor year, whereas the productivity has declined from 2,555 kgs. in 1980-81 to 1,880 kgs. in 1985-86 and increased to 27,508 kgs. in 1990-91 and again declined to 2,656 kgs. in 1996-97 and again increased to 30,896 kgs. in 2006-07. A lot of variation is seen in total subsidies as well as in productivity in Pondicherry, Himachal Pradesh throughout the study period.

The percentage share of Andhra Pradesh in total subsidies has increased from 7.55 in 1980-81 to 8.92 in 2000-01 and percentage share in productivity increased from 3.59 in 1980-81 to 6.13 in 1985-86 and declined to 5.51 in 2006-07. Karnataka has got 5.48 percentage share of subsidies and 4.87 percentage share of productivity in 1980-81. Its percentage share in subsidies declined from 6.21 in 1985-86 to 3.78 in 2006-07, whereas its percentage share in productivity declined from 7.13 in 1985-86 to 5.37 in post-liberalisation period (2006-07).

The percentage share of Kerala in total subsidies declined from 4.13 in 1980-81 to 2.66 in 1996-97 and further declined to 1.36 in 2006-07 and the percentage share in productivity has increased from 1.69 in 1980-81 to 8.88 in 1985-86 and declined to 3.83 in 1990-91 and again increased to 5.66 in 1996-97 and again declined to 4.64 in 2006-07. Tamil Nadu has got first position by receiving maximum amount of subsidies during pre as well as post-liberalisation periods and third, second, first rank in 1980-81, 1985-86 and 2006-07 respectively (See the Table 5.2).

This table reveals that total subsidies in Rs. crores subsidies have increased in absolute terms in all the states of south zone during pre as well post-liberalisation periods, on the other hand, variations are seen in productivity. As compared to the year 2006-07 with the year 1990-91, the above table indicates that subsidies have increased by 9.55 times in Andhra Pradesh, more than six times in Karnataka, 2.66 times in Kerala and 1.8 times in Tamil Nadu, whereas productivity increased maximum 1.33 times in Kerala, 1.29 times in Andhra Pradesh, 1.12 times in Karnataka and 1.03 times in Tamil Nadu. In 1990-91, Andhra Pradesh has increased near about two times of subsidies and 1.22 times of productivity, whereas in 2006-07, Andhra Pradesh has received 6.5 times of subsidies and 1.19 times of productivity as compared to Kerala. During pre- liberalisation period (in 1990-91), Tamil Nadu has got near about two times more of subsidies and 1.36 times of productivity than that of Karnataka, on the other hand, during post-liberalisation period (in 2006-07), Karnataka has got 1.77 times more of subsidies as compared to Tamil Nadu, whereas Tamil Nadu has got 0.79 times of productivity than that of Karnataka.

The total subsidies and productivity in all the states of west zone in India are shown in Table 5.3. This table reveals that Gujarat has got Rs.75.13, Rs.7,254.54 and Rs.4,058.85 in 1980-81,2000-01 and 2006-07 respectively, whereas the productivity has increased from 39,981 kgs. in 1980-81 to 1,11,660 kgs. in 1985-86 and declined to 99,253 kgs. in 1990-91 and again increased to 1,09,041 kgs. in 2006-07. Total subsidies have increased from Rs.12.74 in 1980-81 to Rs.2,415.53 in 2000-01 and declined to Rs.2,187.83 in 2006-07 and productivity has declined from 21455 kgs. in 1980-81 to 44257 kgs. In 1990-91 and increased to 66716 kgs. in 2006-07 in Madhya Pradesh.

**Table 5.3: State-wise Distribution of Total Subsidies and Productivity in West Zone in India during 1980-81 to 2006-07 (Subsidies (Subs.) in Rs./Hectare, Productivity (Prod.) in kgs./Hectare)**

| | West Zone | | | | | | | | | | | |
|---|---|---|---|---|---|---|---|---|---|---|---|---|
| Years | 1980-81 | | 1985-86 | | 1990-91 | | 1996-97 | | 2000-01 | | 2006-07* | |
| Subs. and Prod./Zones | Subs. | Prod. | Subs. | Prod. | Subs. | Prod. | Subs. | Prod. | Subs. | Prod. | Subs. | Prod. |
| Gujarat | 75.13 (5.51) | 39,981 (8.55) | 354.97 (6.81) | 1,11,660 (9.09) | 1,157.54 (8.19) | 99,253 (5.59) | 3,665.34 (9.39) | 1,03,628 (7.12) | 7,254.54 (11.44) | 1,00,993 (5.86) | 4,058.85 (3.45) | 1,09,041 (5.58) |
| Madhya Pradesh | 12.74 (0.94) | 21,455 (4.59) | 78.31 (1.50) | 50,537 (4.12) | 221.66 (1.57) | 44,257 (2.49) | 1,122.54 (2.87) | 59,280 (4.07) | 2,415.53 (3.81) | 56,875 (3.30) | 2,187.83 (1.86) | 66,716 (3.41) |
| Chhattisgarh | – | – | – | – | – | – | – | – | 96.68 (0.15) | 13,529 (0.78) | 4,258.99 (3.62) | 12,426 (0.64) |
| Maharashtra | 68.41 (5.02) | 21,255 (4.55) | 295.78 (5.67) | 1,01,227 (8.25) | 877.78 (6.21) | 99,248 (5.59) | 2,473.10 (6.33) | 92,254 (6.34) | 2,897.57 (4.57) | 93,276 (5.41) | 10,170.16 (8.65) | 87,295 (4.47) |
| Rajasthan | 37.48 (2.75) | 11,171 (2.39) | 154.35 (2.96) | 51,947 (4.23) | 285.56 (2.02) | 74,873 (4.22) | 952.84 (2.44) | 68,565 (4.71) | 1,768.42 (2.79) | 59,463 (3.45) | 673.58 (0.57) | 81,415 (4.17) |
| Goa and Daman & Diu | 25.59 (1.88) | 8,291 (1.77) | 66.03 (.27) | 2,311 (0.19) | 191.13 (1.35) | 53,006 (2.99) | 1,704.71 (4.37) | 4211 (0.29) | 3,268.97 (5.15) | 51,538 (2.99) | 7,720.44 (6.57) | 60,185 (3.08) |
| Dadra Nagar Haveli | – | 1,259 (0.27) | – | 1,682 (0.14) | – | 5,472 (0.31) | 228.48 (0.59) | 1,785 (0.12) | 302.74 (0.48) | 11,962 (0.69) | – | 4,876 (0.25) |

*Source*: (1) Government of Punjab, Statistical Abstract, various years; (2) Government of India, Fertilizers Association, New Delhi; (3) Government of India, State Electricity Boards (SEBs), Annual reports, various years; (4) Government of India, Water Price policy, 2010.

*Note*: (1) * Subsidies of this year does not include irrigation subsidies; (2) Percentages are shown in parentheses.

In Maharashtra, subsidies have increased from Rs.68.41 in 1980-81 to Rs. 10,170.16 in 2006-07, whereas productivity has increased from 21,255 kgs. in 1980-81 to 93,276 kgs. in 2000-01 and declined to 87295 kgs. in 2006-07. Rajasthan received subsidies of Rs.37.48 in 1980-81, which have increased to Rs.952.84 in 1996-97 and declined to Rs.673.58 in 2006-07. On the other hand, productivity has increased from 11171 kgs. in 1980-81 to 74,873 kgs. in 1990-91 and declined to 59,463 kgs. in 2000-01 and again increased to 81,415 kgs. in post-liberalisation period (2006-07).

Goa and Daman and Diu have received Rs.25.59 of subsidies in 1980-81, increased to Rs.7,720.44 in 2006-07 and productivity declined from 8291 kgs. in 1980-81 to 2,311 kgs. in 1985-86 and increased to 60,185 kgs. in 2006-07, whereas in Dadra Nagar Haveli the amount of total subsides have increased from Rs.228.48 in 1996-97 to 302.74 in 2000-01 and declined by 100 per cent in the year 2006-07 (from 2000-01). On the other hand, the productivity has increased from 1,259 kgs. in 1980-81 to 5,472 kgs. in 1990-91 and declined to 1,785 kgs. in 1996-97 and again increased to 11,962 kgs. in 2000-01 and again declined to 4,876 kgs. in 2006-07 in Dadra Nagar Haveli.

During pre as well as post-liberalisation periods, it is found that Gujarat has got first position in subsidies (during 1980-81 to 2000-01) as well as in productivity during 1980-81 to 2006-07. Maharashtra has got second position during 1980-81 to 1996-97 in subsidies and during 1996-97 to 2006-07 in case of productivity, whereas Rajasthan has got third rank in during 1980-81 to 1990-91. Madhya Pradesh achieved fourth position in subsidies and in productivity during pre as well as post-liberalisation periods except in 1980-81 and 2006-07(See the Table 5.3).

It is observed that subsidies have increased in absolute terms in all the states of west zone except in Gujarat and Rajasthan (in these states total subsidies have declined in 2006-07) and variations are observed in case of productivity in all the states throughout the study period. As the year 2006-07 is compared to the year 1990-91, it is found that in Maharashtra, the subsidies have increased the maximum number of times *i.e.* 11.58 and the productivity minimum *i.e.* 0.88 times (which is approximately same in Dadra Nagar Haveli). During post-liberalisation period (in 2006-07), in Gujarat as well as in Rajasthan the productivity has increased in same number of times *i.e.* 1.09, whereas Gujarat has received 3.57 times of more subsidies and Rajasthan 2.36 times of subsidies as compared to pre-liberalisation period (in 1990-91).

The total subsidies and productivity of all the states of north zone in India is shown in Table 5.4. This table reveals that in Haryana, total subsidies have increased from Rs.142.56 in 1980-81 to Rs.3,471.93 in 2006-07. This table further reveals that the productivity has increased from 29,674 kgs. in 1980-81 to 96,561 kgs. in 1990-91 and declined to 83,624 kgs. in 1996-97 and again increased to 1,00,459 kgs. in post-liberalisation period (2006-07). In Punjab, total subsidies have increased from Rs.171.98 in 1980-81 to Rs.21,454.79 in 2006-07 and the productivity of important crops has also increased from 36,270 kgs. in 1980-81 to 92,964 kgs. in 1985-86 and declined to 88,683 kgs. in 1990-91 and again increased to 1,19,883 kgs. in 2000-01 and again declined to 94,434 kgs. in 2006-07. Uttar Pradesh has got Rs.100.55 in 1980-81 as total subsides, increased to Rs.1,868.92 in 2000-01 and declined to Rs.1,799.27 in

**Table 5.4: State-wise Distribution of Total Subsidies and Productivity in North Zone in India during 1980-81 to 2006-07 (Subsidies (Subs.) in Rs./Hectare, Productivity (Prod.) in Kgs./hectare)**

| | North Zone | | | | | | | | | | | |
|---|---|---|---|---|---|---|---|---|---|---|---|---|
| *Years* | *1980-81* | | *1985-86* | | *1990-91* | | *1996-97* | | *2000-01* | | *2006-07** | |
| *Subs. and Prod./Zones* | *Subs.* | *Prod.* | *Subs.* | *Prod.* | *Subs.* | *Prod.* | *Subs.* | *Prod.* | *Subs.* | *Prod.* | *Subs.* | *Prod.* |
| Haryana | 142.56 (10.46) | 29,674 (6.35) | 391.41 (7.50) | 71,026 (5.79) | 970.06 (6.86) | 96,561 (5.44) | 2,521.15 (6.46) | 83,624 (5.74) | 5,135.98 (8.10) | 86516 (5.02) | 3,471.93 (2.95) | 1,00,459 (5.14) |
| Punjab | 171.98 (12.62) | 36,270 (7.76) | 607.74 (11.65) | 92,964 (7.57) | 1513.45 (10.71) | 88,683 (4.99) | 2,898.79 (7.42) | 95,643 (6.57) | 4,854.43 (7.65) | 1,19,883 (6.95) | 21,454.79 (18.24) | 94,434 (4.83) |
| Uttar Pradesh | 100.55 (7.38) | 27,086 (5.79) | 341.13 (6.54) | 66,941 (5.45) | 851.45 (6.02) | 84,660 (4.77) | 1,578.12 (4.04) | 93,162 (6.40) | 1,868.92 (2.95) | 87,936 (5.10) | 1,799.27 (1.53) | 73,211 (3.75) |
| Jammu & Kashmir | 48.58 (3.57) | 14,287 (3) | 226.62 (4.35) | 4,163 (0.34) | 617.34 (4.37) | 53,980 (3.04) | 1,261.73 (3.23) | 4,746 (0.33) | 1,098.24 (1.73) | 17,282 (1.00) | 1,535.92 (1.31) | 8,040 (0.92) |
| Delhi | – | 19,208 (4.11) | – | 3,405 (0.28) | – | 11,017 (0.62) | 1,919.42 (4.91) | 795 (0.05) | 737.57 (1.16) | 2,715 (0.16) | 282.98 (0.24) | 16,396 (0.84) |
| Himachal Pradesh | 25.43 (1.87) | 11,544 (2.47) | 110.02 (2.11) | 7,487 (0.61) | 261.03 (1.85) | 30,588 (1.72) | 257.02 (0.66) | 39,155 (2.69) | 505.62 (0.80) | 14,690 (0.85) | 1,075.89 (0.91) | 6,529 (1.87) |
| Uttaranchal | – | – | – | – | – | – | – | – | – | 68,228 (3.96) | 1,915.29 (1.63) | 66,484 (3.40) |

*Source*: (1) Government of Punjab, Statistical Abstract, various years; (2) Government of India, Fertilizers Association, New Delhi; (3) Government of India, State Electricity Boards (SEBs), Annual reports, various years; (4) Government of India, Water Price policy, 2010.

*Note*: (1) * Subsidies of this year does not include irrigation subsidies; (2) Percentages are shown in parentheses.

2006-07. The productivity has increased from 27,086 kgs. in 1980-81 to 93162 kgs. in 1996-97 and declined 73,211 kgs. in 2006-07 in Uttar Pradesh.

In Jammu & Kashmir, the subsidies have increased from Rs.48.5 in 1980-81 to Rs.1,261.73 in 1996-97 and declined to Rs.1,098.24 in 1996-97 and again increased to Rs.1,535.92 in 2006-07, whereas the productivity has declined from 14,287 kgs. in 1980-81 to 4,163 kgs. in 1985-86 and increased to 53,980 kgs. in 1990-91 and again declined to 4,746 kgs. in 1996-97 and again increased to 18,040 kgs. in 2006-07. Delhi has got Rs.1,919.42 as subsidies in 1996-97, declined to Rs.282.98 in 2006-07, whereas productivity also declined from 19,208 kgs. in 1980-81 to 3,405 kgs. in 1985-86 and increased to 11,017 kgs. in 1990-91 and again declined to 795 kgs. in 1996-97.

This table also shows that for receiving subsidies Punjab has got first rank during pre as well as post-liberalisation periods except in 2000-01 and also ranked first in productivity except in 1990-91 and 2006-07. Haryana achieved second position in subsidies during pre as well as post-liberalisation periods except in 2000-01 and also has got second rank in 1980-81, 1985-86, first rank in 1990-91, 2006-07 and third rank in 1996-97, 2000-01. Uttar Pradesh is on third rank in case of total subsidies during 1980-81 to 2006-07. On the other hand, Jammu & Kashmir has got fourth position by receiving 3.57 per cent of total subsidies in 1980-81, 4.35 per cent in 1985-86, 4.37 per cent in 1990-91, 3.23 per cent in 1996-97, 1.73 per cent in 2000-01 and 1.31 per cent in 2006-07. Uttarakhand Pradesh achieved 1.63 percentage share in subsidies and 3.40 in productivity in 2006-07. Himachal Pradesh has got fifth rank in subsidies, whereas has got a small percentage share in productivity at the centre level during pre as well as post-liberalisation periods (See the Table 5.4).

It is found that the subsidies have increased in absolute terms and variations are seen in case of productivity in all the states of north zone during pre as well as post-liberalisation periods. As post-liberalisation period (in 2006-07) is compared to pre-liberalisation period (in 1990-91), the above table reveals that subsidies have increased in all the states, whereas productivity has also increased in all the states except in Uttar Pradesh and Jammu & Kashmir. In Punjab, subsidies have increased maximum times *i.e.* 14.11, whereas productivity has increased by 1.06 times and Haryana got 3.58 times and 1.04 times more of subsidies and productivity respectively. In 2006-07, Himachal Pradesh has received 19.94 times more of total subsidies and 2.58 times of productivity than that of Himachal Pradesh.

The total subsidies and productivity in sates of east zone of India is shown Table 5.5. This table reveals that in Bihar subsidies have increased from Rs.38.07 in 1980-81 Rs.2,478.49 in 2006-07. Odisha has got Rs.19.66 and Rs. 2724.34 of total subsidies in 1980-81 and 2006-07 respectively. On the other side, variations are seen in productivity *i.e.* the productivity has increased from 19,200 kgs. in 1980-81 to 86,388 kgs. in 1990-91 and declined to 73,262 kgs. in 2000-01 and again increased to 79,981 kgs. in 2006-07.

West Bengal received the maximum amount of subsidies of Rs.57.95 among all the states of this zone in 1980-81, increased to Rs.3,294.53 in 2006-07, whereas productivity has increased from 29,795 kgs. kgs. in 1996-97 to 99,100 kgs. in 2006-07. This table indicates that in Bihar, the percentage share in subsidies has declined

**Table 5.5: State-wise Distribution of Total Subsidies and Productivity in East Zone in India during 1980-81 to 2006-07 (Subsidies (Subs.) in Rs./Hectare, Productivity (Prod.) in Kgs./hectare)**

| | East Zone | | | | | | | | | | | |
|---|---|---|---|---|---|---|---|---|---|---|---|---|
| *Years* | *1980-81* | | *1985-86* | | *1990-91* | | *1996-97* | | *2000-01* | | *2006-07** | |
| *Subs. and Prod./Zones* | *Subs.* | *Prod.* | *Subs.* | *Prod.* | *Subs.* | *Prod.* | *Subs.* | *Prod.* | *Subs.* | *Prod.* | *Subs.* | *Prod.* |
| Subs. and Prod./Zones | | Subs. | Prod. | Subs. | Prod. | Subs. | Prod. | Subs. | Prod. | Subs. | Prod. | Subs. Prod |
| Bihar | 38.07 | 17,481 | 262.16 | 44,918 | 607.95 | 69,464 | 1,014.31 | 62,336 | 1,870.78 | 60,509 | 2,478.49 | 63,783 |
| | (2.79) | (3.74) | (5.03) | (3.66) | (4.30) | (3.91) | (2.60) | (4.28) | (2.95) | (3.51) | (2.11) | (3.260 |
| Jharkhand | – | – | – | – | – | – | – | – | – | – | 1,547.51 | 44,230 |
| | | | | | | | | | | | (1.32) | (2.26) |
| Orissa | 19.66 | 19,200 | 60.30 | 81,675 | 146.60 | 86,388 | 679.20 | 80,601 | 928.49 | 73,262 | 2,724.34 | 79,981 |
| | (1.44) | (4.11) | (1.16) | (6.65) | (1.04) | (4.87) | (1.74) | (5.54) | (1.46) | (4.25) | (2.32) | (4.09) |
| West Bengal | 57.95 | 29,795 | 222.14 | 78,647 | 554.39 | 10,1655 | 1,030.99 | 1,07,509 | 1,731.14 | 1,03,043 | 3,294.53 | 99,100 |
| | (4.25) | (6.37) | (4.26) | (6.41) | (3.92) | (5.72) | (2.64) | (7.38) | (2.73) | (5.97) | (2.80) | (5.07) |

*Source*: (1) Government of Punjab, Statistical Abstract, various years; (2) Government of India, Fertilizers Association, New Delhi; (3) Government of India, State Electricity Boards (SEBs), Annual reports, various years; (4) Government of India, Water Price policy, 2010.

*Note*: (1) * Subsidies of this year does not include irrigation subsidies; (2) Percentages are shown in parentheses.

from 2.79 in 1980-81 to 2.11 in 2006-07. The percentage share of the same state in case of productivity declined from 3.74 kgs. in 1980-81 to 3.66 kgs. in 1985-86 and increased to 4.28 kgs. in 1996-97 and again declined to 3.26 kgs. in 2006-07.

Odisha has got a little percentage share of subsidies *i.e.* 1.44, whereas contributed 4.11 percentage share in productivity in 1980-81. During pre as well as post-liberalisation periods in Odisha, percentage share of subsidies increased from 1.04 in 1990-91 to 1.74 in 1996-97 and declined to 1.46 in 2000-01 and again increased to 2.32 in 2006-07, on the other hand, productivity has increased from 4.87 in 1990-91 to 5.54 in 1996-97 and declined to 4.09 in 2006-07. West Bengal has got first rank in subsidies as well as productivity during per-liberalisation period (1980-81, See the Table 5.5).

In east zone, it is observed that in all the states subsidies have increased and variations are found in productivity in absolute terms during 1980-81 to 2006-07. As post-liberalisation period (in 2006-07) is compared to pre-liberalisation period (in 1990-91), it is found that in Odisha the subsidies have increased maximum *i.e.* 18.58 times followed by West Bengal (5.94 times) and Bihar (4.07 times), whereas productivity has declined in Bihar 1.09 times, followed by Odisha (1.08 times) and West Bengal (1.02 times). In 2006-07, West Bengal has received 1.6 times more of productivity and 1.32 times more of subsidies as compared to Bihar. In 1990-91, West Bengal has got 1.46 times of more productivity than that of Bihar and Bihar has got 1.09 times of more subsidies than that of West Bengal.

The total subsidies and productivity of all the states of north-east zone of India is shown in Table 5.6. This table indicates that Assam state has got subsidies of Rs.25.59 in 1980-81, increased to Rs.165.81 and declined to Rs.159.44 in 1996-97 and again increased to 837.17 in 2006-07 and productivity of this state also increased from 12,804 kgs. in 1980-81 to 54,182 kgs. in 1990-91 and declined to 49,377 kgs. in 2006-07. In Tripura, total subsidies have increased from Rs.8.91 in 1980-81 to Rs.330.37 in 1990-91 and declined to Rs.186.16 in 1996-97 and again increased to 962.96 in 2006-07, whereas productivity in the same state has declined from 22,244 kgs. in 1980-81 to 20,737 kgs. in 1996-97.

In Manipur, the total subsidies has increased from Rs.49.04 in 1980-81 to 2,325.77 in 1996-97 and declined to Rs.2,176.17 in 2000-01 and again increased to Rs.10,988.83 in 2006-07. Productivity has declined from 12,475 kgs. in 1980-81 to 45,829 kgs. in 2006-07. In Meghalaya, the subsidies have increased from Rs.10.26 in 1980-81 to Rs.137.68 in 1996-97 and declined by 68.49 per cent in 2006-07 as compared to 2000-01, on the other side, variations are seen in productivity throughout the study period.

During study period, it has been found that in Nagaland the total subsidies have increased from Rs.0.45 in 1980-81 to Rs.19.56 in 1990-91 and declined to Rs.11.36 in 2000-01 and again increased to Rs.11.40 in 2006-07, whereas the productivity has declined from 10,592 kgs. in 1980-81 to 1,903 kgs. in 1985-86 and increased to 46,277 kgs. in 1990-91 and again declined to 64,061 kgs. in 2006-07. Mizoram received a little amount of subsidies *i.e.* Rs.0.58 in 1980-81, increased to Rs.203.71 in 2006-07, whereas in productivity variations are seen.

**Table 5.6: State-wise Distribution of Total Subsidies and Productivity in North-East Zone in India during 1980-81 to 2006-07 (Subsidies (Subs.) in Rs./Hectare, Productivity (Prod.) in kgs./Hectare)**

| | North-East Zone | | | | | | | | | | | |
|---|---|---|---|---|---|---|---|---|---|---|---|---|
| Years | 1980-81 | | 1985-86 | | 1990-91 | | 1996-97 | | 2000-01 | | 2006-07* | |
| Subs. and Prod./Zones | Subs. | Prod. | Subs. | Prod. | Subs. | Prod. | Subs. | Prod. | Subs. | Prod. | Subs. | Prod. |
| Assam | 25.59 (1.88) | 12,804 (2.74) | 69.15 (1.33) | 52,849 (4.30) | 165.81 (1.17) | 54,182 (3.05) | 159.44 (0.41) | 53,093 (3.65) | 487.50 (0.77) | 50,645 (2.94) | 837.17 (0.71) | 49,377 (2.53) |
| Tripura | 8.91 (0.65) | 22,244 (4.76) | 104.53 (2.00) | 1,398 (0.11) | 330.37 (2.34) | 70,911 (3.99) | 186.16 (0.48) | 20,737 (1.42) | 344.00 (0.54) | 20,737 (1.20) | 962.96 (0.82) | 68,551 (3.51) |
| Manipur | 49.04 (3.60) | 12,475 (2.67) | 224.39 (4.30) | 1,583 (0.13) | 1,080.63 (7.65) | 42,377 (2.39) | 2,325.77 (5.96) | 39,429 (2.71) | 2,176.17 (3.43) | 49,972 (2.900 | 10,988.83 (9.34) | 45,829 (2.35) |
| Meghalaya | 10.26 (0.75) | 14,475 (3.10) | 29.72 (0.57) | 10,257 (0.84) | 40.35 (0.29) | 32,638 (1.84) | 137.68 (0.35) | 11,113 (0.76) | 180.21 (0.28) | 14,557 (0.84) | 56.78 (0.05) | 18,114 (0.93) |
| Nagaland | 0.45 (0.03) | 10,592 (2.27) | 2.66 (0.05) | 1,903 (0.16) | 19.56 (0.14) | 46,277 (2.61) | 17.62 (0.05) | 8,988 (0.62) | 11.36 (0.02) | 75,251 (4.36) | 11.40 (0.01) | 64,061 (3.28) |
| Mizoram | 0.58 (0.04) | 6,930 (1.48) | 116.75 (2.24) | 769 (0.06) | 415.82 (2.94) | 9,646 (0.54) | 18.25 (0.05) | 1,719 (0.12) | 127.03 (0.20) | 12,861 (0.75) | 203.71 (0.17) | 14,128 (0.72) |
| Sikkim | 4.06 (0.30) | | 18.81 (0.36) | 3,346 (0.27) | 37.24 (0.26) | 12,981 (0.73) | 26.94 (0.07) | 7,368 (0.51) | 70.35 (0.11) | 9,509 (0.55) | 26.42 (0.02) | 9,823 (0.50) |
| Arunchal Pradesh | – | 3,679 (0.79) | – | 2,178 (0.18) | – | 10,849 (0.61) | 12.17 (0.03) | 9,863 (0.68) | 19.33 (0.03) | 10,569 (0.61) | 31.62 (0.03) | 22,866 (1.17) |

*Source*: (1) Government of Punjab, Statistical Abstract, various years; (2) Government of India, Fertilizers Association, New Delhi; (3) Government of India, State Electricity Boards (SEBs), Annual reports, various years; (4) Government of India, Water Price policy, 2010.

*Note*: (1) *Subsidies of this year does not include irrigation subsidies; (2) Percentages are shown in parentheses.

In Sikkim, the amount of total subsidies have increased from Rs.4.06 in 1980-81 to Rs.37.24 in 1990-91 and declined to Rs.26.94 in 1996-97 and again increased to Rs.70.35 in 2000-01 and again declined to Rs.26.42 in 2006-07, whereas the productivity within the same state has increased from 3,346 kgs. in 1980-81 to 12,981 kgs. in 1985-86 and declined to 7,368 kgs. in 1990-91 and again increased to 9,823 kgs. in 2006-07.

It is found that in Assam the percentage share of subsidies has declined from 1.88 in 1980-81 to 0.41 in 1996-97 and increased to 0.77 in 2000-01 and declined to 0.71 in 2006-07, whereas percentage share in productivity has increased from 2.74 in 1980-81 to 4.30 in 1985-86 and declined to 2.53 in 2006-07. The percentage share of subsidies in Tripura is less than its percentage share in productivity during 1980-81 to 2006-07, except in 1985-86, same pattern is found in Meghalaya, Nagaland, whereas in Manipur, Mizoram, Sikkim and in Arunachal Pradesh, the percentage share of subsidies is more than that of their productivity during pre as well as post-liberalisation periods (See the Table 5.6).

This table indicates that in absolute terms, variations are found in the subsidies as well productivity in all the states of north-east zone during the study period. As compared the year 2006-07 with the year 1990-91, it is observed that subsidies have increased in all the states except in Nagaland, Mizoram and Sikkim and productivity has increased except in Meghalaya and Sikkim. During 1996-97 to 2000-01, in Tripura productivity remains same (20737 kgs.), whereas in 2000-01, total subsidies have increased by 1.85 times more as compared to 1996-97.

# Chapter 6

# Summary, Conclusions and Policy Implications

The socio-economic structure, which prevailed prior to the British rule in the country, resulted in the organization of self-sufficient villages. It has been maintaining some kind of static equilibrium. The Indian peasant, though not properly educated, has adequate experience of farming systems and he has been dependent on it for the means of living. The Royal commission of Agriculture in India observed that both the methods of cultivation and social organization exhibit that settled order which is characteristic of all countries in which the cultivating peasant has long lived in and closely adapted himself to the conditions of a particular environment.

The Indian agrarian economy on the eve of independence was critical in situation. It could be characterized totally primitive, deteriorative and turbulent. After partition, the country is left with 82 per cent of the total population of undivided India as well as only with 69 per cent of land under rice, 65 per cent under wheat and 75 per cent under all cereals. The deficiency of food grains is quite alarming and aggravating at that time (Chahal, 1999).

In view of this, after independence tremendous efforts are made to boost the economy through agriculture as one of the tools for development. The Government of India adopted a more positive approach and hence a well defined policy of integrated production programmes with defined targets and a proper distribution programme is adopted along with other measures for the overall economic development of the country. Specific programmes like new agriculture technology are introduced to convert agriculture into a successful and prosperous business, to bring more land under cultivation and to raise agriculture production. In India, the adoption of new agricultural technique is costly than that of traditional method of cultivation. In traditional method, inputs are least expensive, on the other hand, inputs in modern technology like high yielding varieties of seeds, fertilizers, farm mechanization and irrigation are very costly and Indian farmers being poor are not in a position to buy these expensive inputs. Then on the recommendations of food grain price committee (Jha Committee), the Government of India started the scheme of subsidies on purchase of various agriculture inputs to facilitate the farmers (Singh, 1994).

Subsidies have occupied agricultural economists for a long time because they are pervasive in agriculture, even though they are often applied in ways that benefit mostly richer farmers, cause inefficiencies, lead to a heavy fiscal burden, distort trade, and have negative environmental effects. Agricultural subsidies can play an important role in early phases of agricultural development by addressing market failures and promoting new technologies (Fan, 2008).

All of these subsidies by reducing the prices of the inputs, served in the initial stages of green revolution, as incentives to the farmers for adopting the newly introduced seed-cum-fertilizer technology. These helped in raising the agricultural output, after some time, the amount paid on these subsidies began to rise (Gulati, 2003).

In India, at present centre as well as state governments are providing subsidies on fertilizers, irrigation (canal water), electricity and other subsidies to marginal farmers and farmers' cooperative societies in the form of seeds, development of oil seeds, pulses, cotton, rice, maize and crop insurance schemes and price support schemes etc. Out of these subsidies, the Central Government of India provides indirect subsidies to farmers on the purchase of fertilizers from 1977, whereas state governments are providing subsidies on irrigation as well as on electricity (Government of Punjab, Agriculture Department, Chandigarh).

Subsidies are often criticized for their financial burden. Some researchers assert to the extent that these should be withdrawn in a phased manner, such a step will reduce the fiscal deficit, improve the efficiency of resources use, funds for public investment in agriculture. On the other hand, there is a fear that agriculture production and income of farmers would decline if subsidies are curtailed. These are very important issues and need serious investigation. The objectives of the present study are to study the growth and distribution of agricultural subsidies in India, to study the growth and distribution of agricultural subsidies in Punjab State, to study the impact of agricultural subsidies in Punjab, to suggest ways and means for giving agricultural subsidies to farmers of Punjab.

The present study is related to agriculture subsidies in India from 1980-81 to 2008-09. In this study agriculture subsidies of fertilizers, electricity, irrigation (canal water), seeds, machinery etc. are discussed during pre-liberalisation period (during 1980-81 to 1985-86), first phase of liberalisation period (during 1990-91 to 1996-97) as well as during second phase of liberalisation period (during 2000-01 to 2008-09). For analysing the growth and distribution pattern of agriculture subsidies, five zones *i.e.* south zone (includes Andhra Pradesh, Karnataka, Kerala, Tamil Nadu, Pondicherry, Andaman and Nicobar Islands and Lakshadweep), west zone (includes Gujarat, Madhya Pradesh, Chhattisgarh, Maharashtra, Rajasthan, Goa, Daman and Diu and

Dadra Nagar Haveli), east zone (Bihar, Jharkhand, Odisha and West Bengal), north zone (Haryana, Punjab, Uttar Pradesh, Uttarakhand, Himachal Pradesh, Jammu & Kashmir, Delhi and Chandigarh) and north-east zone (Assam, Tripura, Manipur, Meghalaya, Nagaland, Arunachal Pradesh, Mizoram and Sikkim) have been taken.

The present study has been divided into six chapters. The first chapter provides an introduction to the concept of agriculture subsidies. The second chapter is related to the different views of the analysts. The third chapter deals with the gross cropped area and fourth chapter includes fertilizers subsidies, electricity subsidy and irrigation subsidy in India. The fifth chapter deals the issues relating to the productivity of crops and total subsidies in India. The last chapter presents the summary, conclusions and policy implications.

The present study is related to agriculture subsidies in India from 1980-81 to 2008-09. In this study, agriculture subsidies of fertilizers, electricity, irrigation, seeds, machinery etc. and impact of agriculture subsidies are discussed during pre-liberalisation period (1980-81 to 1985-86), first phase of liberalisation period (1990-91 to 1996-97) as well as during second phase of liberalisation period (2000-01 to 2008-09). Following are the major findings of the present study:

The gross cropped area (GCA in thousand hectares), has increased by only 1.01 times during post-liberalisation period (in 2006-07) as compared to pre-liberalisation period (in 1980-81) in India, whereas declined by 1.06 times in the year 2006-07 as compared to 1990-91. In absolute terms, in north-east zone, the gross cropped area has increased throughout the study period, declined in south zone (in 2000-01) and also declined in west as well as in north zones in 2006-07. Whereas in east zone, the GCA has increased during 1980-81 to 1990-91 and declined during 1996-97 to 2006-07. The percentage-wise analysis shows that west zone has occupied first rank followed by north, south, east and east-north zones during pre as well as post-liberalisation periods.

During post-liberalisation period (in 2006-07), the GCA (in 000 hectares) in absolute terms has declined in all the states of south zone except in Karnataka and Tamil Nadu (in these states, the GCA has increased) as compared to pre-liberalisation period (in 1990-91). The percentage share has increased in all the states except in Pondicherry, Lakshadweep (percentage share remains constant) and in Andaman and Nicobar (percentage share declined) in the year 2006-07 as compared to the year 1990-91. In all the states of west zone, a lot of variation is observed in absolute terms as well as in percentage share of GCA during the study period.

The GCA, in absolute terms, has increased in Haryana, Punjab, Uttar Pradesh and Jammu & Kashmir states of north zone during 1980-81 to 2000-01 and declined in Delhi during 1980-81 to 2006-07, whereas a lot of variation is seen in Himachal Pradesh. In Haryana, Uttar Pradesh, Jammu Kashmir and in Himachal Pradesh, it has increased by 1.08 times, 1.01 times, 1.05 times and 1.02 times respectively in 2006-07 as compared to 1990-91. During pre-liberalisation period (1990-91), Punjab has received 7.04 times more of GCA as compared to Jammu & Kashmir and Himachal Pradesh has got thirteen times more from Delhi. Whereas during post-liberalisation period (2006-07), Punjab has got near about four times from Jammu & Kashmir and Himachal Pradesh received twenty two times than that of Delhi. It is found that Uttar Pradesh has got 4.04 times and 4.3 times more of GCA as compared to Haryana during pre-liberalisation period (in 1990-91) and post-liberalisation period (in 2006-07). During pre-liberalisation in Bihar state of east zone, GCA has declined in absolute terms, on the other hand, it has increased in Odisha as well as in West Bengal. In all the states of north-east zone, the GCA in absolute terms has increased only in Nagaland during pre as well as post-liberalisation periods, whereas a lot of variation is seen in percentage share among all the other states.

At national level during pre as well as post-liberalisation periods, the total subsidies (fertilizers, electricity and irrigation subsidies) in Rs. crores have increased at different increasing rates (as compared to other years, the rate is high during pre-liberalisation period *i.e.*, 290.39 per cent in 1985-86 from 1980-81) and in absolute terms. In 2008-09, the total subsidies have increased 94.38 times more than that of 1980-81, whereas fertilizers subsidies have increased by 29 times, electricity subsidy by 75.24 times and irrigation subsidy by 36.86 times in 2000-01 as compared to pre-liberalisation period (1980-81). In 2000-01, irrigation subsidies have increased by 1.07 times more, whereas electricity subsidy has increased by 1.96 times more as compared to fertilizers subsidies. All the subsidies in Rs. crores, have increased during pre and post-liberalisation periods, whereas during 1980-81 to 1985-86 the percentage share of subsidies of fertilizers in total subsidies is maximum followed by irrigation and electricity. During 1996-97 to 2000-01, the percentage share of electricity in total subsidies is maximum followed by irrigation and fertilizers, whereas in 2008-09, the percentage share of fertilizers subsidies is again maximum *i.e.* 87.26 per cent.

The north zone has got major percentage share (36.49 per cent, 32.2 per cent in 1980-81 and 1985-86 respectively) of total subsidies at national level. Whereas during 1990-91 to 2000-01, west zone occupied topmost position and again north zone is ahead among all the other zones by getting 31.80 per cent of total subsidies at national level in 2008-09. The percentage share of east zone in total subsidies has increased during the study period except in 1990-91 and a lot of variation is seen in percentage share of north-east zone throughout the study period.

During pre as well as post-liberalisation periods, the total subsidies in Rs. crores, have increased in absolute terms in all the states of south zone. In all the states of west zone, the total subsidies in Rs. crores, including fertilizers, electricity and irrigation subsidies have increased in absolute terms, whereas percentage-wise analysis shows a lot of variation during the study period. The total subsidies in Rs. crores, have increased in absolute terms in all the states of north zone except in Jammu & Kashmir and Delhi. As compared to 2008-09 to 1990-91, it is observed that Haryana is that state, where total subsidies are increased maximum number of times *i.e.* 13.6, whereas only 6.5 times in Jammu & Kashmir. As compared to Haryana, Punjab has got 1.99 times more of total subsidies in 1990-91.

Total subsidies in Rs. crores, have increased in absolute terms in all the states of east zone during pre as well post-liberalisation periods, whereas a lot of variation is seen in percentage share. In absolute terms, the total subsidies in Rs. crores, have increased in all the states of north-east zone except in Tripura, Manipur, Mizoram and Sikkim throughout the study period.

During pre-liberalisation period (in 1990-91), it is observed that Uttar Pradesh is leading among all the other states of India by receiving maximum percentage share of gross cropped area *i.e.* 13.74 followed by Madhya Pradesh (12.88 per cent), Maharashtra (11.79 per cent), Rajasthan (10.45 per cent) and Andhra Pradesh and in post-liberalisation period (in 2006-07), Rajasthan is ahead among all the other states by getting 20.67 per cent of GCA followed by Uttar Pradesh (14.69 per cent), Andhra Pradesh and Gujarat (7.29 per cent), Karnataka and Madhya Pradesh (7.08 per cent) and Tamil Nadu (4.64 per cent). In total subsidies (in Rs. crores), Uttar Pradesh has ranked first, Maharashtra, Andhra Pradesh, Gujarat, Rajasthan and Punjab ranked second, third, fourth, fifth and sixth respectively in 1990-91, whereas Uttar Pradesh leading among all the states followed by Andhra Pradesh, Maharashtra, Punjab and Karnataka in 2008-09.

At national level as well as zone level, fertilizers subsidies in Rs. crores, have increased in absolute terms and a lot of variation is seen in percentage share during pre as well as post-liberalisation periods. The fertilizers subsidies have increased the maximum times *i.e.* more than forty five times in north-east zone. As compared to east zone, during pre-liberalisation period (in 1990-91), the south zone has got 2.28 times more of fertilizers subsidies, whereas in 2008-09, it has received 1.81 times more. It is seen that during pre-liberalisation period, north-zone received maximum (40.47 per cent in 1980-81 and 40.76 per cent in 1985-86) amount of fertilizers subsidies followed by south, west, east and north-east zones. In 1996-97, south zone has occupied first rank by getting 33.73 per cent followed by north, west, east and north-east zones, whereas in 2005-06, North zone has got 31.79 per cent followed by west, south, east and north-east zones.

In all the states of south zone, the fertilizers subsidies in Rs. crores, have increased in absolute terms during the study period. As the year 2008-09 is compared to the year 1990-91, it is found that in Karnataka the fertilizers subsidies have increased twenty four times, in Andhra Pradesh near about twenty one times and in Kerala near about twelve times. Andhra Pradesh has got near about two times fertilizers subsidies as compared to Karnataka during pre as well as post-liberalisation periods. In all the states of west zone, the fertilizers subsidies in Rs. crores, in absolute terms have increased during post-liberalisation period as compared to pre-liberalisation period. Whereas a lot of variation is seen in percentage share during the study period. As the year 2008-09 is compared to the year 1990-91 of pre-liberalisation period, it is found that in Rajasthan, the fertilizers subsidies have increased 30.97 times, in Gujarat 26.55 times, in Madhya Pradesh more than nineteen times and in Maharashtra, these have risen up more than twenty one times.

The fertilizers subsidies in Rs. crores, have increased in absolute terms in all the states of north zone (except in Delhi) during the pre as well post-liberalisation periods, whereas a lot of variation is seen in percentage-wise analysis. As 2008-09 is compared to 1990-91, it is found that in Uttar Pradesh, these have increased near about 19.65 times more of fertilizers subsidies, Jammu & Kashmir 26.95 times, Haryana twenty four times, Himachal Pradesh eighteen times and Punjab fifteen times. In 1990-91, Punjab has got two times more of fertilizers subsidies and 1.37 times in 2008-09 than that of Haryana, whereas Uttar Pradesh has received 52.61 times more in 1990-91 and 38.37 times in 2008-09 from Jammu & Kashmir. In all the states of east zone as well as north-east zone, the fertilizers subsides in Rs. crores, have been increased in absolute terms during the study period.

In fertilizers subsidies (in Rs. crores), Uttar Pradesh is ahead among all the by getting maximum percentage share 17.96 followed by Andhra Pradesh (12.98 per cent), Maharashtra (10.56 per cent), Punjab (9.79 per cent) and Madhya Pradesh (7.77 per cent) in 1990-91, on the other hand in 2008-09, Uttar Pradesh has got first position by receiving 16.19 per cent of fertilizers subsidies, followed by Andhra Pradesh (12.33 per cent), Maharashtra (10.30 per cent), Karnataka (7.35 per cent) and Punjab (7.10 per cent).

In India as well as in the five zones the fertilizers subsidises in Rs. per hectare, have increased in absolutes terms during pre as well as post-liberalisation periods. As post-liberalisation period (2006-07) is compared to pre liberalisation period (1990-91), it is observed that in south, the fertilizers subsidies have increased by fourteen times, in east ten times, west zone more than nine times, north zone more than six times and north-east zone six times. As compared to west zone, south zone has got near about two times of fertilizers subsidies in 1990-91 and more two times in 2006-07, whereas north zone has got near about three times in 1990-91 and 1.66 times in 2006-07 as compared to east zone. North zone has enjoyed major portion of fertilizers subsidies per hectare at national level during 1980-81 to 1996-97, whereas south zone has got maximum percentage share *i.e.* 45.97 in 2000-01 and 43.64 in 2006-07. It is found that in all the states of south zone the fertilizers subsidies in Rs. per hectare have increased in absolutes terms during 1980-81 to 2006-07, whereas the percentage share has increased in Andhra Pradesh, Karnataka and Kerala during 1980-81 to 1990-91 and a lot of variation is observed during post-liberalisation period.

The fertilizers subsidies in Rs. per hectare have been increased in the entire states (except in Goa) of the west zone as well as north zone throughout the study period, whereas percentage share analysis reveals a lot of variations. While 2006-07 is compared to 1990-91, in Maharashtra these have increased the maximum *i.e.* fourteen times, whereas in Goa it has increased minimum *i.e.* 2.11 times. In all the states of east zone, the fertilizers subsidies in Rs. per hectare have increased in absolute terms during pre as well as post-liberalisation periods. The analysis of fertilizers subsidies in Rs. per hectare shows that in all the states of north-east zone these have increased except in Meghalaya, Nagaland, Mizoram and Sikkim in absolute terms during the study period. As 2006-07 is compared to 1990-91, it is observed that in Assam these have risen up by eighteen times, in Tripura more than eight times, in Manipur near about six times, in Mizoram as well as in Arunachal Pradesh near about five and Meghalaya only 1.4 times.

The electricity subsidy in Rs. crores has increased in absolute terms during the study period except in 2008-09, whereas the same pattern is found in south, west and east zones. As 2008-09 is compared to 1990-91, in north zone this subsidy has increased by 4.68 times more, in south zone more than three times, in east zone more than two times and in west zone 2.02 times. The results indicate that north zone occupied first rank by getting maximum amount of 47.34 per cent and 31.91 per cent of electricity subsidies in Rs. crores in 1980-81 and 1985-86 respectively. West zone has got first rank after liberalisation period except in 2008-09. East and north-east zone achieved fourth and fifth rank during pre as well as post-liberalisation periods.

In absolute terms, the electricity subsidy in Rs. crores, has increased in all the states of south zones during 1980-81 to 2000-01. While comparing the year 2008-09 with the year 1990-91, in Kerala this has increased more than eighty four times, in Karnataka more than four times and in Tamil Nadu only four times. As compared to Kerala, in 1990-91, Karnataka has got 37.43 times more of electricity subsidy, Tamil Nadu has got 51.32 times, on the other hand in 2008-09, in Karnataka, this has risen up by near about two times more and in Tamil Nadu more than two times. In all the states of west zone, the electricity subsidy in Rs. crores, has increased in absolute terms during pre as well as post-liberalisation periods except in 2008-09. As post-liberalisation period (2008-09) is compared to pre-liberalisation period (1990-91), in Rajasthan this has increased near about six times, in Madhya Pradesh more than four times and in Gujarat only two times. Gujarat has got 2.2 times more of electricity subsidy and 1.2 times in 1990-91 and 2008-09 respectively as compared to Madhya Pradesh.

In Haryana as well as in Punjab states of north zone, electricity subsidy in Rs. crores, has increased in absolute terms during pre and post-liberalisation periods, whereas in Uttar Pradesh and in Jammu & Kashmir, this subsidy has declined in 2000-01. As the year 2008-09 is compared to the year 1990-91, it is found that in Haryana, power subsidy has increased by near about twelve times more, in Punjab more than five and in Uttar Pradesh more than two times. In Punjab, this has increased more than two times in 1990-91 as compared to Haryana, whereas in 2008-09, Haryana has got 1.01 times more of electricity subsidy as compared to Punjab.

In absolute terms, the electricity subsidy in Rs. crores, has increased in all the states of east zone throughout the study period. As post-liberalisation period (1996-97) is compared to pre liberalisation period (1980-81), it is found that in West Bengal this has risen up by more than two hundred fourteen times, in Odisha more than fifty and in Bihar near about nineteen times. In north-east zone, the electricity subsidy in Rs. crores, has increased during pre and post-liberalisation periods, whereas the increasing rate is higher in 1990-91 among all the other years of study.

In 1990-91, Uttar Pradesh has ranked first by receiving maximum electricity subsidy (in Rs. crores) *i.e.* 15.21 per cent followed by Maharashtra (12.79 per cent), Gujarat (11.67 per cent), Punjab (10.81 per cent) and Andhra Pradesh (10.41 per cent), whereas Haryana, which has got 17.85 per cent of electricity subsidy, is leading the other states of India like Punjab (17.61 per cent), Tamil Nadu (12.40 per cent), Uttar Pradesh (10.37 per cent) and Karnataka (10.09 per cent) during post-liberalisation period (in 2008-09).

At national as well as zones level the electricity subsidy in Rs. per hectare, has increased in absolute terms during pre as well as post-liberalisation periods. As the year 2000-01 is compared to the year 1990-91, it is found that in India it has increased by near about six times, in west zone nine times, south zone more than six times, in north zone near about four times, in east zone 3.4 times and north-east zone more than two times. North zone has got approximately five times more of electricity subsidy and near about seven times as compared to east in 1990-91 and 2000-01 respectively. At India level, electricity subsidies in Rs. per hectares are Rs. 343.26 per hectare, out of which north zone has got 44.39 per cent followed by south zone (34.16 per cent), west zone (15.61 per cent), east zone (5.57 per cent) and north-east zone (0.26 per cent) in 1980-81. In 1990-91, these subsidies are Rs. 4,083.80 per hectare in India, out of which most of percentage share has gone to north zone (36.57 per cent) followed by south, west, east and north-east zones. It is observed that in absolute terms, the electricity subsidy in Rs. per hectare, has increased in all the states (except in Madhya Pradesh) of south zone as well west zone during the study period.

In all the states of north zone except in Jammu & Kashmir and Himachal Pradesh, the electricity subsidy in Rs. per hectare, has increased in absolute terms, whereas in case of percentage share a lot of variation is seen during pre as well as post-liberalisation periods. As the year 2000-01 is compared to the year 1990-91, in Haryana, this has increased by more than eight times, in Punjab 4.26 times. In all the states of east zone in absolute terms, electricity subsidy in Rs. per hectare has increased during the study period.

At national level as well as zone level the irrigation subsidy in Rs. crores, has increased in absolute terms during pre as well post-liberalisation periods. As post-liberalisation period (2006-07) is compared with post-liberalisation (1990-91), it is found at national level this has increased near about nine times, whereas in south zone this has increased near about nineteen times, in west zone has got more than seven times, in east as well as in north-east zones near about five and in north zone more than four times of irrigation subsidy. In 1990-91, west zone has got subsidy more than two times and north zone near about three times more of irrigation subsidy as compared to south and east respectively, whereas in post-liberalisation period (in 2006-07), south zone has received 1.23 times and north zone near about three times more of irrigation subsidy as compared to west zone and east zone respectively. Out of five zones of India, west zone is ahead among all the other states during pre as well as post-liberalisation periods except in 1980-81 and in 2006-07 (in these two years south zone has got first rank).

In absolute terms, the irrigation subsidy in Rs. crores has increased in all the states of south zone during the study period. In all the states of west zone except in Madhya Pradesh, the irrigation subsidy in Rs. crores has risen up in absolute terms during pre as well as post-liberalisation periods. As post-liberalisation period (in 2006-07) is compared to pre-liberalisation period (in 1990-91), in Gujarat this has increased approximately nine times and in Maharashtra, Rajasthan and Madhya Pradesh received near about seven times. In 1990-91, Gujarat has got approximately 1.95 times more of irrigation subsidy as compared to Madhya Pradesh and Maharashtra received more than four times as compared to Rajasthan.

In all the states of north zone irrigation subsidy in Rs. crores, has increased except in Punjab, Jammu & Kashmir and Himachal Pradesh during pre as well as post-liberalisation periods. As the year 2006-07 is compared to the year 1990-91, it is observed that in Haryana, this has increased near about seven times more of irrigation subsidy, in Uttar Pradesh as well as in Himachal Pradesh more than four times, in Jammu Kashmir 2.12 two times and in Punjab 3.48 times. In all the states (except in West Bengal) of east zone the irrigation subsidy in Rs. crores, has increased in absolute terms throughout the study period, whereas a lot of variation is seen in all the states of north-east zone during the study period.

Maharashtra has got topmost position by getting 21.44 per cent of irrigation subsidy (in Rs. crores), Uttar Pradesh (16.27 per cent), Gujarat (10.18 per cent), Andhra Pradesh (9.46 per cent) and Karnataka (5.53 per cent) has got second, third, fourth and fifth position in pre-liberalisation period (in 1990-91). Whereas in post-liberalisation period (in 2006-07), it is observed that Andhra Pradesh (30.62 per cent), followed by Maharashtra (16.37 per cent), Karnataka (11.00 per cent), Gujarat (10.30 per cent) and Uttar Pradesh (8.20 per cent).

Irrigation subsidy in Rs. per hectare shows that at India level as well as zone level this subsidy has risen up during pre as well as post-liberalisation periods. South zone is ahead by getting maximum percentage share 36.24 in 1980-81, north zone is ahead among all the other zones by getting 27.20 per cent in 1985-86, whereas north-east zone is leading the other zones by obtaining 35.01 per cent in 1990-91 and again west zone has occupied topmost position by receiving 28.79 per cent in 1996-97 and 35.55 per cent in 2000-01. Irrigation subsidy in Rs. per hectare in all the states except in Tamil Nadu of south zone has increased in absolute terms throughout the study period. In absolute terms, the irrigation subsidy in Rs. per hectare has increased in all states of west zone during pre as well as post-liberalisation periods. As the year 2006-07 is compared to the year 1990-91, it is observed that in Madhya Pradesh, this has increased more than twelve times and in Gujarat 7.2 times. It is found that in 1990-91, Gujarat has got more than four times, whereas in 2006-07, 2.63 times as compared to Madhya Pradesh. In 1996-97, Goa has received 4.09 times and in 2006-07 near about twenty times than that of Rajasthan.

As post-liberalisation period (2006-07) is compared to pre-liberalisation period (1990-91), it is observed that in Punjab, this has increased the maximum *i.e.* more than sixty five times, in Haryana more than six times and in Himachal Pradesh more than four times. In 1990-91, Uttar Pradesh has got 1.1 times more of irrigation subsidy than that of Haryana, whereas in 2006-07, this has increased by approximately thirteen times as compared to Uttar Pradesh. On the other hand, Jammu & Kashmir has got three times and 1.36 times in 1990-91 and 2006-07 as compared to Himachal Pradesh. During pre as well as post-liberalisation periods, irrigation subsidy in Rs. per hectare has been increased in all the states of east zone in absolute terms, whereas a lot of variation is seen in percentage-wise analysis.

A lot of variation is observed in absolute terms as well as in percentage-wise analysis of irrigation subsidy in Rs. per hectare in all the states of north-east zone during the study period. As post-liberalisation period (2006-07) is compared to pre-liberalisation period (1990-91), in Manipur this has risen up in the maximum ten times, in Assam as well as in Tripura 1.44 only. Manipur has got approximately eight times more, whereas 60.22 times in 1990-91 and 2006-07 respectively as compared to Assam.

As GCA is compared to total subsidies, it is observed that at national level, the increasing rate of total subsidies (fertilizers, electricity and irrigation) is higher than gross cropped area (GCA) during pre, first as well as second phase of liberalization periods. In 1985-86, the total subsidies increased by 290.39 per cent and GCA 2.4 as compared from 1980-81, in 1996-97 total subsidies increased by 174.78 per cent and GCA 1.73 as from 1990-91, whereas total subsidies increased by 109.53 per cent in 2008-09 and GCA declined by 5.84 per cent in 2006-07 as compared to 2000-01 at the national level.

There is a lot of variation to find out the relationship between gross cropped area (GCA) and in total subsidies in zones as well as in states throughout the study period. As zone level, it is observed that there is a negative relationship between GCA and total subsidies, in west zone and in north zone (in 2006-07) and in east zone (in 1996-97 and in 2006-07) and at state level in Andhra Pradesh (in 1985-86 and 2006-07), Karnataka (in 2000-01), Kerala, Punjab, West Bengal and Assam (in 2006-07), Tamil Nadu (during 1990-91 to 2000-01), Gujarat and Mizoram (in 1985-86), Madhya Pradesh (in 2006-07), Maharashtra (in 1996-97), Rajasthan and Tripura (in 2000-01), Bihar (during 1985-86 to 2006-07), Odisha (during 1996-97 to 2006-07), Manipur (during 1985-86 to 1990-91), the GCA has declined, whereas total subsidies have increased.

It is seen that there is a direct relationship in GCA and total subsidies *i.e.* GCA as well as total subsidies have increased at zone level in west and in north (during 1980-81 to 2000-01), in south zone during 1980-81 to 1996-97 and in north-east zone (during 1980-81 to 2006-07) and at state level, in Andhra Pradesh, Gujarat and West Bengal (in 1990-91), Tamil Nadu (in 1985-86), Karnataka, Haryana (1985-86 and 2006-07), Madhya Pradesh, Rajasthan (in 1996-97), Maharashtra (in 2000-01), Uttar Pradesh, Jammu & Kashmir and in Odisha (in 1985-86).

During pre as well as post-liberalisation periods, at country level as well as zone level, the total subsidies (in Rs. per hectare) have increased in absolute terms, whereas at India level as well as in south, west, north, north-east zones, productivity (in kgs per hectare) has also increased except in 1996-97 and in east zone productivity has declined during 1996-97 to 2000-01. As compared to post-liberalisation period (in 2006-07) with pre-liberalisation period (in 1990-91), it is observed that in India, subsidies have increased 8.32 times, whereas productivity increased by only 1.1 times. While comparing the same time period, as zone level analysis shows that in west zone, subsidies have increased the maximum number of times *i.e.* 11.95 times, followed by south zone (8.93 times), east zone (7.67 times), north zone (7.49 times) and north-east zone (6.28 times), On the other hand productivity has increased maximum *i.e.* 1.90 times in south zone, followed by west zone (1.12 times), north zone (1.11 times), east zone (1.1 times) and north-east zone (1.05 times). In 1990-91, south zone has got near about three times of total subsidies and has near about two times of productivity, whereas in 2006-07, it has received 3.37 times of subsidies and has near about two times of productivity as compared to east zone.

Total subsidies in Rs. per hectare have increased in absolute terms in all the states of south zone during pre as well post-liberalisation periods, on the other hand, variations are seen in productivity in kgs per hectare. As compared to the year 2006-07 with the year 1990-91, subsidies have increased by 9.55 times in Andhra Pradesh, more than six times in Karnataka, 2.66 times in Kerala and 1.8 times in Tamil Nadu, whereas productivity increased maximum 1.33 times in Kerala, 1.29 times in Andhra Pradesh, 1.12 times in Karnataka and 1.03 times in Tamil Nadu. In 1990-91, Andhra Pradesh has increased near about two times of subsidies and 1.22 times of productivity, whereas in 2006-07, Andhra Pradesh has received 6.5 times of subsidies and 1.19 times of productivity as compared to Kerala. During pre- liberalisation period (in 1990-91), Tamil Nadu has got near about two times of subsidies and 1.36 times of productivity than that of Karnataka, on the other hand, during post-liberalisation period (in 2006-07), Karnataka has got 1.77 times of subsidies as compared to Tamil Nadu, whereas Tamil Nadu has got 0.79 times of productivity than that of Karnataka. Subsidies in Rs. per hectare have increased in absolute terms in all the states of west zone except in Gujarat and Rajasthan (in these states total subsidies have declined in 2006-07) and variations are observed in case of productivity (in kgs per hectare) in all the states throughout the study period.

It is found that the total subsidies have in Rs. per hectare increased in absolute terms and variations are seen in case of productivity (in kgs per hectare) in all the states of north zone during pre as well as post-liberalisation periods. As post-liberalisation period (2006-07) is compared to pre-liberalisation period (1990-91), the subsidies have increased in all the states, whereas productivity has also increased in all the states except in Uttar Pradesh and Jammu & Kashmir. In Punjab, subsidies have increased maximum times *i.e.* 14.11, whereas productivity has increased by 1.06 times and Haryana has got 3.58 times more and 1.04 times of subsidies and productivity respectively. In 2006-07, Himachal Pradesh has received 19.94 times more of total subsidies and 2.58 times of productivity than that of Himachal Pradesh.

In east zone, it is observed that in all the states total subsidies in Rs. per hectare have increased and variations are found in productivity (in kgs per hectare) in absolute terms during 1980-81 to 2006-07. On the other hand, variations are found in the total subsidies in Rs. per hectare as well productivity (in kgs per hectare) in all the states of north-east zone during the study period.

State-wise analysis shows that states like Tamil Nadu (in 1980-81), Pondicherry (during 1996-97 to 2006-07), Gujarat (in 2000-01), Maharashtra (in 2006-07) and Mizoram (during 1985-86 to 1990-91) have received more amounts of total subsidies at national level, these have got less percentage share of productivity, while some states like Kerala (during 1996-97 to 2006-07), Rajasthan Karnataka, Tamil Nadu (in 2006-07), Madhya Pradesh (during 1980-81 to 1996-97), Odisha (during 1980-81 to 2006-07) and Gujarat (in 1980-81) have showed better performance in case of productivity by consuming a little amount of subsidies.

As state-wise analysis shows that during pre-liberalisation period (in 1990-91), Tamil Nadu has ranked first by getting maximum percentage share of productivity (in Kgs. per hectare) *i.e.* 7.21, followed by West Bengal (5.72 per cent), Gujarat (5.59 per cent), Maharashtra (5.58 per cent) and Haryana (5.44 per cent), whereas during post-liberalisation (2006-07), Tamil Nadu (6.73 per cent) has got first position followed Gujarat (5.58 per cent), Andhra Pradesh (5.51 per cent), Haryana (5.14 per cent) and West Bengal (5.07 per cent).

## Policy Implications and Suggestions

Most of the studies either supported distributing subsidies or withdrawal of subsidies. However, the present study reveals that some subsidies should be given and some others can be withdrawn without harming the farmers. Withdrawal of subsidies should be carried out in phased manner. Following are the some suggestions emerging out of the present study:

The centre government should adopt some criteria to give away subsidies to states either on the basis of gross cropped area or productivity.

From the study it has been noted that subsidies which have direct relationship on productivity and income like seeds, fertilizers should be given to farmers, on the other hand, subsidies on electricity can be withdrawn as supply of electricity in Punjab is irregular moreover farmers prefer regular supply of power even if they have to pay for it. If implemented, it will reduce state electricity board's burden and this amount can be used for production of more electricity, reducing the need of purchasing electricity at very high prices, which adds to the deficit of state finance.

Due to irregular supply of electricity and canal water, farmers have to use diesel pump sets to irrigate the crops. The expenditure of diesel pump sets is very high as compared to flat rates of electricity and nominal charges of canal water. From farmers point of view they are ready to pay bills for irrigation as uninterrupted supply of canal water and electricity is given to agriculture sector. As a result government should impose flat rates on electricity supply given to agriculture sector.

Subsidy should be given to those areas where, it is actually required like if water table depletion is to be checked then farmer should be motivated to grow crops like potatoes, oilseeds and sunflowers etc., subsidies should be given to promote these practices, it will reduce quantity of fertilizers and water usage for cultivation. Checking of water table depletion will reduce expenditure on higher horse power motors viz-a-viz subsidy on electricity. Provision of market facilities will also be helpful in motivating farmers to grow these crops.

Unbalanced fertilizer use does not lead to immediately visible harmful effects whereas it adversely affects soil quality over time. Excessive use of insecticides also affects the human health. Farmers should be educated about the effects of unbalance use of fertilizers and they should be encouraged to use these in proper ratio. Government should provide training to the farmers for increasing the productivity of crops. Different types of campaigns should be organised at village level to encourage them to use environment-friendly practices and balanced use of chemicals, fertilizers, increasing area under high yielding varieties of seeds etc.

Government should formulate farmer friendly agriculture price policy, under which the price of farm produce should be fixed keeping in view the rising costs of farm inputs; this will help in making the farmers financially independent.

In view of drought/deficit rainfall in certain regions (Bihar, Jharkhand, Odisha and West Bengal), it was decided by centre government to implement a diesel subsidy during kharif (in 2010) to save standing crops in the field, same pattern should be followed in states where this problem occurs.

Government should keep aside its motive to please voters or strengthen the vote bank, it should frame rational policy in which small size category farmers, who are not actual beneficiaries of subsidies, could get more and subsides, which they do not want should be withdrawn.

The cooperative sector should be developed/strengthened in the state. The sale of fertilizers, high yielding varieties of seeds should be promoted through village level primary agriculture co-operative societies. It will help in checking diversion of loans as well as use of spurious chemicals purchased from commission agents.

Government should provide the funds timely to Punjab State Electricity Board to (Now it is unbundled into Punjab State Power Cooperation Ltd. and Punjab State Transmission Cooperation Ltd.) improve the conditions of the Power Plants and to recruit new man-power in order to improve performance of the electricity department.

Subsidies should be given to those who actually need, like small and medium size category farmers. Subsidies, which they do not need should be withdrawn but in a phased manner. On the other hand, instead of subsidies, the government should focus on just three things - electricity generation, infrastructural development and water supply. The accompanying development will take care of the rest. The subsidies should be replaced with constructive schemes that empower people and give them that one push they need to get out of poverty.

# Appendices

**Table A.1: Availability, Imports and Subsidies of Fertilisers in India during 1980-81 to 2008-09**

| *Years* | *Availability of Fertiliser (000 tonnes)* | | | | *Total Fertiliser Subsidy (Rs. In Crores)* | | | *Rate of Fertiliser Subsidy (Rs. per tonne)* | | | *Rate of Fertiliser Subsidy (Rs. Per kg.)* | | |
|---|---|---|---|---|---|---|---|---|---|---|---|---|---|
| | *Domestic* | *Imports* | *Total* | *Percentage of Imports in Total Fertiliser* | *Domestic Subsidy* | *Import Subsidy* | *Total Subsidy* | *Domestic Subsidy* | *Import Subsidy* | *Total Subsidy* | *Domestic Subsidy* | *Import Subsidy* | *Total Subsidy* |
| 1980-81 | 3005 | 2759 | 5764 | 47.86607 | 170 | 335 | 505 | 565.72 | 1214.21 | 876.13 | 0.5657 | 1.2142 | 0.876 |
| 1985-86 | 5756 | 3399 | 9155 | 37.12725 | 1600 | 324 | 1924 | 2779.71 | 953.22 | 2101.58 | 2.7797 | 0.9532 | 2.101 |
| 1990-91 | 9045 | 2758 | 11803 | 23.36694 | 3730 | 659 | 4389 | 4123.83 | 2389.41 | 3718.55 | 4.1238 | 2.3894 | 3.718 |
| 1995-96 | 11335 | 3955 | 15290 | 25.86658 | 4300 | 1935 | 6235 | 3793.56 | 4892.54 | 4077.83 | 3.7935 | 4.8925 | 4.077 |
| 1996-97 | 11155 | 2026 | 13181 | 15.37061 | 5415 | 1350 | 6765 | 4854.33 | 6663.38 | 5132.39 | 4.8543 | 6.6633 | 5.132 |
| 2000-01 | 14704 | 2091 | 16795 | 12.45013 | 13799 | 1 | 13800 | 9384.52 | 4.78240 | 8216.73 | 9.3845 | 0.0047 | 8.216 |
| 2005-06 | 15575 | 5253 | 20828 | 25.22086 | 16249 | 2050 | 18299 | 10432.74 | 3902.53 | 8785.77 | 10.432 | 3.9025 | 8.785 |
| 2006-07 | 16095 | 6080 | 22175 | 27.41826 | 22001 | 3950 | 25951 | 13669.46 | 6496.71 | 11702.82 | 13.669 | 6.4967 | 11.702 |
| 2007-08 | 14707 | 7583 | 22290 | 34.01974 | 33538 | 6800 | 40338 | 22804.11 | 8967.43 | 18096.9 | 22.804 | 8.9674 | 18.096 |
| 2008-09 | 14334 | 10151 | 24485 | 41.45804 | 66858 | 32598 | 99456 | 46642.95 | 32113.09 | 40619.15 | 46.642 | 32.113 | 40.619 |

*Source*: Government of India, Economic Survey of Various Years.

**Table A.2: State-wise Distribution of Fertilizers Consumption in South Zone in India during 1980-81 to 2008-09 (In 000 tonnes)**

| *Years/States* | *South Zone* | | | | | | | |
|---|---|---|---|---|---|---|---|---|
| | *1980-81* | *1985-86* | *1990-91* | *1996-97* | *2000-01* | *2005-06* | *2006-07* | *2008-09* |
| Andhra Pradesh | 574.6 | 888.1 | 1619.75 | 1768.79 | 2174.571 | 2552.55 | 2484.1 | 3070.9 |
| Karnataka | 343.9 | 555.6 | 832.91 | 825.88 | 1348.353 | 1524.92 | 1485.8 | 1831.8 |
| Kerala | 97.5 | 142.6 | 244.38 | 187.57 | 173.205 | 202.46 | 208.92 | 260.91 |
| Tamil Nadu | 491.3 | 668.3 | 830.94 | 790.99 | 963.002 | 1099.22 | 1124.96 | 1265.22 |
| Pondicherry | – | – | – | 22.23 | 23.677 | 42.58 | 44.14 | 27.94 |
| Andaman and Nicobar Islands | – | – | – | 0.42 | 0.42 | 0.59 | 0.61 | 0.6 |
| Lakshadweep | – | – | – | 0.16 | – | – | – | – |

*Source*: Government of India, Fertilizers Association, Fertilizer Statistics, various issues, New Delhi.

**Table A.3: State-wise Distribution of Fertilizers Consumption in West Zone in India during 1980-81 to 2008-09 (In 000 tonnes)**

| *Years/States* | *West Zone* | | | | | | | |
|---|---|---|---|---|---|---|---|---|
| | *1980-81* | *1985-86* | *1990-91* | *1996-97* | *2000-01* | *2005-06* | *2006-07* | *2008-09* |
| Gujarat | 356.9 | 421.3 | 706.39 | 813.54 | 750.64 | 1279.92 | 1408.79 | 1716.98 |
| Madhya Pradesh | 196.8 | 431.9 | 812.36 | 973.11 | 960.759 | 940.85 | 1205.11 | 1423.4 |
| Chhattisgarh | – | – | – | – | – | 374.32 | 436.23 | 462.82 |
| Maharashtra | 421 | 669.1 | 1317.36 | 1329.34 | 1647.1 | 1968.38 | 2258.97 | 2566.11 |
| Rajasthan | 135.1 | 220.9 | 371.02 | 664.809 | 664.809 | 882.06 | 936.13 | 1052.02 |
| Goa | 4.1 | 7.2 | 7.8 | 6 | 5.843 | 5.52 | 5.95 | 8.13 |
| Daman and Diu | – | – | – | 0.36 | 0.26 | – | 0.49 | 0.39 |
| Dadra Nagar Haveli | – | – | – | 1.12 | 0.923 | 1.13 | 1.25 | 1.1 |

*Source*: Government of India, Fertilizers Association, Fertilizer Statistics, various issues, New Delhi.

**Table A.4: State-wise Distribution of Fertilizers Consumption in North Zone in India during 1980-81 to 2008-09 (In 000 tonnes)**

| *Years/States* | *North Zone* | | | | | | | |
|---|---|---|---|---|---|---|---|---|
| | *1980-81* | *1985-86* | *1990-91* | *1996-97* | *2000-01* | *2005-06* | *2006-07* | *2008-09* |
| Haryana | 230.3 | 372 | 586.3 | 761.46 | 930.295 | 1129 | 1125 | 1289 |
| Punjab | 753.6 | 1098.2 | 1197.81 | 1207.73 | 1313.636 | 1687.15 | 1691 | 1768 |
| Uttar Pradesh | 1150.6 | 1971.9 | 2240.91 | 2768.76 | 3068.342 | 3464.26 | 3726.08 | 4032.76 |
| Jammu & Kashmir | 20.7 | 35.2 | 42.59 | 43.33 | 64.98 | 92.26 | 86.03 | 105.1 |
| Delhi | – | – | – | 23.71 | 5.127 | 0.45 | 1.04 | 0.64 |
| Himachal Pradesh | 16.2 | 23.6 | 34.6 | 34.45 | 35.552 | 47.97 | 48.98 | 57.37 |
| Uttarakhand | – | – | – | – | – | 12.07 | 143.14 | 152.83 |

*Source*: Government of India, Fertilizers Association, Fertilizer Statistics, various issues, New Delhi.

**Table A.5: State-wise Distribution of Fertilizers Consumption in East Zone in India during 1980-81 to 2008-09 (In 000 tonnes)**

| *Years/States* | *East Zone* | | | | | | | |
|---|---|---|---|---|---|---|---|---|
| | *1980-81* | *1985-86* | *1990-91* | *1996-97* | *2000-01* | *2005-06* | *2006-07* | *2008-09* |
| Bihar | 204.5 | 501.5 | 598.46 | 785.83 | 987.16 | 918.96 | 1072 | 1357 |
| Jharkhand | – | – | – | – | – | 134.45 | 138.51 | 147.27 |
| Odisha | 76.4 | 140.5 | 192.67 | 250.77 | 319.214 | 394.88 | 402.9 | 534.9 |
| West Bengal | 282.8 | 408.7 | 753 | 896.1 | 1085.086 | 1,239.68 | 1365.2 | 1519.3 |

*Source*: Government of India, Fertilizers Association, Fertilizer Statistics, various issues, New Delhi.

**Table A.6: State-wise Distribution of Fertilizers Consumption in North-East Zone in India during 1980-81 to 2008-09 (In 000 tonnes)**

| Years/States | North-East Zone | | | | | | | |
|---|---|---|---|---|---|---|---|---|
| | *1980-81* | *1985-86* | *1990-91* | *1996-97* | *2000-01* | *2005-06* | *2006-07* | *2008-09* |
| Assam | 9.3 | 16.7 | 37.67 | 55.81 | 140.619 | 198.12 | 203.68 | 220.61 |
| Tripura | 1.98 | 5.15 | 9.22 | 8.62 | 9.199 | 15.17 | 203.68 | 220.61 |
| Manipur | 3 | 4.84 | 8.22 | 13.81 | 22.039 | 14.20 | 19.07 | 12.87 |
| Meghalaya | 2.6 | 3 | 2.65 | 3.43 | 3.864 | 4.91 | 5.13 | 3.69 |
| Nagaland | 0.08 | 0.25 | 1.11 | 0.85 | 0.399 | 0.62 | 0.64 | 0.91 |
| Mizoram | 0.07 | 0.09 | 0.83 | 0.39 | 1.444 | 2.28 | 3.7 | 3.7 |
| Sikkim | 0.42 | 1.2 | 1.53 | 0.75 | 1.081 | 0.35 | 0.35 | – |

*Source*: Government of India, Fertilizers Association, Fertilizer Statistics, various issues, New Delhi.

**Table A.7: State-wise Distribution of Expenditure and Receipt in South Zone in India during 1980-81 to 2006-07 (In Rs. Crores)**

| *Years/States* | *South Zone* | | | | | | | | | | | |
|---|---|---|---|---|---|---|---|---|---|---|---|---|
| | *1980-81* | | *1985-86* | | *1990-91* | | *1996-97* | | *2000-01* | | *2006-07* | |
| | *Expenditure* | *Receipts* | *Expenditure* | *Receipts* | *Expenditure* | *Receipts* | *Expenditure* | *Receipts* | *Expenditure* | *Receipts* | *Expenditure* | *Receipts* |
| Andhra Pradesh | 60.48 | 1.18 | 202.75 | 14.35 | 419.56 | 48.29 | 1369.86 | 64.77 | 2210.98 | 11.43 | 10626 | 68.81 |
| Karnataka | 51.89 | 3.94 | 145.52 | 4.29 | 233.31 | 16.89 | 1266.57 | 17.1 | 1750.23 | 18.46 | 3813.93 | 21.48 |
| Kerala | 8.36 | 1.42 | 23.82 | 1.13 | 64.30 | 2.07 | 1249.47 | 175.69 | 1731.77 | 183.68 | 3792.45 | 227.24 |
| Tamil Nadu | 26.73 | 1.86 | 57.41 | 1.29 | 122.98 | 2.11 | 205.46 | 4.64 | 618.12 | 9.31 | 679.86 | 28.51 |
| Pondicherry | – | – | – | – | – | – | 0.55 | 0 | 1.19 | 0 | – | – |

*Source*: Government of India, Pricing of Water in Public System, 2010, Combined Finance and Revenue Accounts of different states.

**Table A.8: State-wise Distribution of Expenditure and Receipt in West Zone in India during 1980-81 to 2006-07 (In Rs. Crores)**

| *Years/States* | *West Zone* | | | | | | | | | | | |
|---|---|---|---|---|---|---|---|---|---|---|---|---|
| | *1980-81* | | *1985-86* | | *1990-91* | | *1996-97* | | *2000-01* | | *2006-07* | |
| | *Expenditure* | *Receipts* | *Expenditure* | *Receipts* | *Expenditure* | *Receipts* | *Expenditure* | *Receipts* | *Expenditure* | *Receipts* | *Expenditure* | *Receipts* |
| Gujarat | 30.45 | 5.63 | 195.75 | 7.32 | 418.70 | 20.01 | 1767.88 | 37.54 | 2699.16 | 136.58 | 3882.13 | 330.62 |
| Madhya Pradesh | 7.71 | 3.79 | 70.30 | 14.4 | 130.42 | 15.84 | 498.14 | 44.71 | 434.28 | 39.55 | 1341.06 | 29.82 |
| Chhattisgarh | – | – | – | – | – | – | – | – | 61.88 | 10.38 | 459.96 | 104.96 |
| Maharashtra | 78.04 | 10.21 | 272.41 | 16.53 | 641.74 | 18.17 | 2196.11 | 58 | 2474.74 | 62.49 | 6088.58 | 444.93 |
| Rajasthan | 39.46 | 6.35 | 170.96 | 12.45 | 220.23 | 16.31 | 797.8 | 24.6 | 882.72 | 18.43 | 14168.3 | 479.06 |
| Goa | – | – | – | – | – | – | 29.95 | 0.36 | 53.5 | 0.22 | 134.83 | 2.93 |
| Dadra Nagar Haveli | – | – | – | – | – | – | – | – | – | – | – | – |

*Source*: Government of India, Pricing of Water in Public System, 2010, Combined Finance and Revenue Accounts of different states.

**Table A.9: State-wise Distribution of Expenditure and Receipt in North Zone in India during 1980-81 to 2006-07 (In Rs. Crores)**

| *Years/States* | *North Zone* | | | | | | | | | | | |
|---|---|---|---|---|---|---|---|---|---|---|---|---|
| | *1980-81* | | *1985-86* | | *1990-91* | | *1996-97* | | *2000-01* | | *2006-07* | |
| | *Expenditure* | *Receipts* | *Expenditure* | *Receipts* | *Expenditure* | *Receipts* | *Expenditure* | *Receipts* | *Expenditure* | *Receipts* | *Expenditure* | *Receipts* |
| Haryana | 39.59 | 4.56 | 82.76 | 12.09 | 148.98 | 17.31 | 397.3 | 24.3 | 536.11 | 54.3 | 990.89 | 87.19 |
| Punjab | 18.02 | 8.47 | 79.98 | 11.40 | 125.25 | 14.16 | 658.98 | 10.69 | 543.05 | 16.33 | 7463.06 | 368.3 |
| Uttar Pradesh | 69.23 | 28.86 | 335.82 | 108.30 | 682.40 | 35.03 | 1050.17 | 100.78 | 1565.83 | 282.13 | 2974.63 | 148.63 |
| Jammu & Kashmir | 3.16 | 0.25 | 16.44 | 0.79 | 34.48 | 0.43 | 21.32 | 0.53 | 39.61 | 0.44 | 82.3 | 1.01 |
| Himachal Pradesh | 0.19 | 0 | 4.48 | 0 | 15.64 | 0.02 | 5.77 | 0 | 15.8 | 0.02 | 44.71 | 0.13 |
| Uttarakhand | – | – | – | – | – | – | – | – | 27.88 | 2.23 | 243.11 | 5.69 |

*Source*: Government of India, Pricing of Water in Public System, 2010, Combined Finance and Revenue Accounts of different states.

**Table A.10: State-wise Distribution of Expenditure and Receipt in East Zone in India during 1980-81 to 2006-07 (In Rs. Crores)**

| *Years/States* | *East Zone* | | | | | | | | | | | |
|---|---|---|---|---|---|---|---|---|---|---|---|---|
| | *1980-81* | | *1985-86* | | *1990-91* | | *1996-97* | | *2000-01* | | *2006-07* | |
| | *Expenditure* | *Receipts* | *Expenditure* | *Receipts* | *Expenditure* | *Receipts* | *Expenditure* | *Receipts* | *Expenditure* | *Receipts* | *Expenditure* | *Receipts* |
| Bihar | 13.82 | 7.73 | 87.81 | 9.62 | 181.83 | 0 | 314.48 | 37.64 | 517.29 | 33 | 637.85 | 12.9 |
| Jharkhand | | | | | | | | | 59.43 | 5.66 | 298.04 | 51.09 |
| Odisha | 12.75 | 3.23 | 28.61 | 4.76 | 61.46 | 4.45 | 396.68 | 6.54 | 488.42 | 18.71 | 741.65 | 49.75 |
| West Bengal | 18.68 | 0.61 | 84.52 | 1.16 | 166.08 | 1.55 | 222.75 | 2.79 | 302.5 | 3.99 | 304.35 | 6.95 |

*Source*: Government of India, Pricing of Water in Public System, 2010, Combined Finance and Revenue Accounts of different states.

**Table A.11: State-wise Distribution of Expenditure and Receipt in North-East Zone in India during 1980-81 to 2006-07 (In Rs. Crores)**

| *Years/States* | *North-East Zone* | | | | | | | | | | | |
|---|---|---|---|---|---|---|---|---|---|---|---|---|
| | *1980-81* | | *1985-86* | | *1990-91* | | *1996-97* | | *2000-01* | | *2006-07* | |
| | *Expenditure* | *Receipts* | *Expenditure* | *Receipts* | *Expenditure* | *Receipts* | *Expenditure* | *Receipts* | *Expenditure* | *Receipts* | *Expenditure* | *Receipts* |
| Assam | 7.83 | 0.14 | 22.06 | 0.41 | 644.09 | 0.26 | 28.08 | 0.07 | 69.01 | 0.15 | 59.52 | 0.38 |
| Tripura | 0.19 | 0.03 | 3.34 | 0 | 11.30 | 0.01 | 7.22 | 0 | 5.66 | 0 | 17.74 | 0 |
| Manipur | 0.89 | 0.08 | 10.07 | 0.16 | 6.54 | 0.40 | 40.17 | 0 | 27.72 | 0.31 | 239.38 | 7.85 |
| Meghalaya | 0.07 | 0 | 1.27 | 0 | 4.25 | 0 | 1.61 | 0 | 2.49 | 0 | 0 | 0 |
| Nagaland | – | – | – | – | 4.56 | 0 | – | – | 12.3 | 0 | – | – |
| Mizoram | – | – | 0.81 | 0 | 3.14 | 0 | – | – | 0.01 | 0 | 0.01 | 0 |
| Sikkim | 0.13 | 0 | 1.90 | 0 | 0 | 0 | – | – | – | – | – | – |
| Arunachal Pradesh | – | – | – | – | 9.13 | 0 | 0.48 | 0 | 0.34 | 0 | 0.56 | 0 |

*Source*: Government of India, Pricing of Water in Public System, 2010, Combined Finance and Revenue Accounts of different states.

# Bibliography


Acharaya, S.S.(1998), "Input subsidies in Indian Agriculture: some issues, policies for agricultural development: perspectives from states", *Economic and Political Weekly*, Vol. 2, No. 3, pp. 143-146, Nov. 4.

Acharaya S.S., Jogi R.L. (2004), "Farm input subsidies in Indian Agriculture", *Agricultural Economic Research Review*, Vol.17, No.1, pp. 120-125, June 5.

Aggarwal, Bina (1983), "*Mechanisation in Indian Agriculture- An analytical study based on the Punjab*", Allied Publishers Private Limited, New Delhi.

Ahluwalia, Sanjeev S. (2000), "Power Tariff Reform in India", *Economic and Political Weekly*, Vol. 35, No.38, pp. 16-22, July 11.

Anderson, Kym and Will, Martin (2007), "Incidence of trade and subsidy policies on developing country welfare, exports and debt substantially", *Economic and Political Weekly*, Vol. 8, No. 3, pp. 13-20, June 3.

Audinet, Pierre (2002), "Electricity prices in India", International Energy Agency, Vol. 1, No. 2, pp. 10-14, March 4.

Asha, P. (1986), "Trends in growth and pattern of subsidies in budgetary operations of central government", Reserve Bank of India, Occasional Papers, Vol. 7, No. 2, pp. 208-250.

Ashra, Sunil and Malini Chakravarty (2007), "Input Subsidies to Agriculture: Case of Subsidies to Fertilizer Industry across Countries", *The Journal of Business Perspective*, Vol. 11, No. 11, pp. 35-58, July 23.

Badiani, Reena and Katrina K. Jessoe (2010), "Electricity Subsidies, Elections, Groundwater Extraction and Industrial Growth in India", *Economic and Political Weekly*, Vol. 3, No. 4, pp. 30-40, March 19.

Badiani, Reena and Katrina K. Jessoe (2011), "Elections at what cost? The Impact of electricity Subsidies on Groundwater Extraction and Agricultural Production", *Economic and Political Weekly*, Vol. 2, No. 3, pp. 41-45, January 23.

Badiani, Reena and Katrina K. Jessoe (2011), "Electricity Subsidies for Agriculture: Evaluating the Impact and Persistence of these Subsidies in India", *Economic and Political Weekly*, Vol.4, No.5, pp. 203-210, September 22.

Bajpai, A.D.N. and S.K. Shivastava (1991), "Relevance of subsidies in determining fertilizer consumption in Indian agriculture-An economic analysis", *Journal of Rural Development*, Vol. 10, No. 4, pp. 391-403, May 4.

Bala, Brij, and R.K. Sharma (2005), "Factors influencing fertilizers production and consumption in India", *Indian Journal of Agricultural Research*", Vol. 39, No. 2, pp. 146-149.

Barman, Kiran (2005), "The problem of budgetary subsidies in India", *Economic Affairs*, Vol. 27, No. 4-6, pp. 371-372, April-June.

Bathla, Seema (1997), "Ground Water Sustainability: A case study of Ludhiana District", *Indian Journal of Agricultural Economic*, Vol. 4, No. 6, pp. 56-60, April 3.

Bhalla, G. S. (2004), "Globalisation and Indian Agriculture, State of the Indian Farmer: A Millennium Study", *Economic and Political Weekly*, Volume 1, No. 4, pp. 70-73, Dec 12.

Bhalla, G.S. and Gurmail Singh (2009), "Economic Liberalisation and Indian Agriculture: A State-wise Analysis", *Economic and Political Weekly*, December, Vol. 25, No. 52, pp. 78-84, Jan 4.

Birner, Regina, Surupa Gupta and Neeru Sharma (2011), "The Political Economy of Agricultural Policy Reform in India: Fertilizers and electricity for irrigation", IFPRI (International Food Policy Research Institute), pp. 56-61, Dec 12.

Chander, Satish (2005), "India's Fertiliser Policy and Scope for Reforms", *Economic and Political Weekly*, Vol. 2, No. 1, pp. 31-42, March 3.

Chandrashekhar, G. (2006), "Should India demand farm subsidy cuts by developed nations?" *Economic and Political Weekly*, Vol. 2, No. 4, pp. 38-51, March 11.

Chattopadhyay, P. (2004), "Cross Subsidy in Electricity Tariff: Evidence from India", *Energy Policy*, Vol. 32, No. 5, pp. 673-684, July 23.

Dass, Nabagopal (2007), "*Agriculture in India - Past, Present and Future*", Deep and Deep Publishers, New Delhi.

De Janvry, A., and K. Subbarao (1984), "Agricultural Price Policy and Income Distribution in India." *Economic and Political Weekly*, Vol. 3, A166-A178, December 19.

Dillion, G.S. (2008), "Subsidies yes, but no freebies", *Indian Express*, June 18.

Dubash, N K and S C Rajan (2001), "Power Politics: Process of Power Sector Reforms in India", *Economic and Political Weekly*, Vol.36, No. 35, pp. 3367-3390, Sep. 1.

Elinder, Liselotte Schafer (2005), "Obesity, hunger and agriculture: The damaging role of subsidies", *Economic and Political Weekly*, Vol. 2, pp. 1333-1336, Dec 3.

Fan, Shengyen, Ashok Gulati and Sukhaseo Thorat (2008), "Investment, Subsidies and Pro-poor growth in Rural India", *Agricultural Economic*, Vol. 39, No. 2, pp. 163-170, Sept. 13.

Fukumi, Atsushi (2007), "The Effects of Political Instability on Power Subsidies: An Analysis of Indian States", *Economic and Political Weekly*, Vol. 2, No. 3, pp. 134-140, Nov. 23.

Government of India, Central Statistical Organisation, New Delhi.

Government of India, Economic Survey, Various years.

Government of India, Fertilizer Association, fertilizer statistics, various issues, New Delhi.

Government of India (1979), Report of committee on control and subsidies, Vol. 1 to 3, Ministry of Finance, New Delhi.

Government of India, Union Budget, Various years (http:/indiabudget.nic.in).

Government of India, Pricing of Water in Public System in India 2010.

Government of Punjab, Annual Report on the Working of State Electricity Boards and Electricity Departments of different states of India, various years.

Government of Punjab, Economic Survey, various years.

Government of Punjab, Statistical Abstract, various years.

Government of India, Annual report on the working of state electricity boards and electricity department, Power and Energy Division, Planning Commission, various issues.

Grossman, Nick and Dylan Carlson (2011), *"Agriculture Policy in India: The Role of Input Subsidies"*, USITC Executive Briefings on Trade, pp. 45-50, Jan 23.

Gupta, Anjali (1984), "Impact of agricultural subsidies", *Economic and Political Weekly*, Vol. 39, No. 4, pp. 48-53, May 18.

Gupta, R.K. (2007), *"Agricultural Subsidies and their Economic Implications"*, Deep and Deep Publishers, New Delhi.

Gulati, Ashok (1989), "Input Subsidies in Indian Agriculture: A State-wise Analysis", *Economic and Political Weekly*, Vol. 24, No. 25, pp. A57-A65, June 24.

Gulati, Ashok (2007), "Investment, subsidies and pro-poor growth in rural India", *Economic and Political Weekly*, volume 18, No. 3, July 19.

Gulati, Ashok and Anil Sharma (1995), "Subsidy Syndrome in Indian Agriculture", *Economic and Political Weekly*, Vol. 30, No. 39, pp. A93-A102, Sept. 30.

Gulati, Ashok and Sudha Narayanan (2003), *"The subsidy syndrome in Indian Agriculture"*, Oxford University Press, New Delhi.

H. John (2006), "Should agricultural subsidies be banned?", *Economic and Political Weekly*, volume 27, No. 5, pp. 71-78, April 23.

Halmandage, B.V. and Dr.N.N.Munde (2010), "A Study of fertilizer subsidy in India", *International Research Journal* Vol. 1, No. 7, pp. 45-50, Dec. 12.

Honnihal, Siddharth (2010), "Estimating Power Consumption in Agriculture", *Economic and Political Weekly*, Vol.39, No.8, pp. 790-792, August 8.

Howes, S and R Murgai (2003), "Incidence of Agricultural Power Subsidies: An Estimate", *Economic and Political Weekly*, Vol. 38, No. 16, pp. 1533-1535, Sept 11.

Jain, Varinder (2006), "Political Economy of Electricity Subsidy: Evidence from Punjab", *Economic and Political Weekly*, Vol. 4, No. 3, pp. 89-92, Sept. 23.

Jakhar, Sunil (2008), "Don't cripple the poor with sops empower them", *The Tribune*, March 19.

Jha, Reghbendra (2007), "Investment and Subsidies in Indian Agriculture", *Economic and Political Weekly*", Vol 1, No. 2, pp. 13-23, August 17.

Joshi, P.K. and A.K. Agnihotri (1982), "Impact of input subsidy on income and equity under land reclamation", *Indian Journal of Agriculture Economic*, Vol. 37, No. 3, pp. 252-260, July-Sept.

Karnik, Ajit and Mala Lalvani (1996)," Interest groups, Subsidies and Public Goods: Farm Lobby in Indian Agriculture", *Economic and Political Weekly*, Vol.31, No.13 pp. 818-820, March 30.

Kaushik, Viren and Uttam Gupta (1999), " Fertilizes subsidy imbroglio : Decontrol the only way out," *Economic and Political Weekly*, Vol. 13, No. 2, pp. 24-31, April 8.

Kaushik, Viren and Uttam Gupta (2003), "Fertilizers Subsidy imbroglio: Decontrol the only way out", *Economic and Political Weekly*, Vol. 10, No. 4, pp. 67-70, May 08.

Khatttar, R. K., C. R. Kaushik and S. D. Chamola (1992), "Extent of input subsidies and their beneficiaries in Indian agriculture", *Indian Journal of Agriculture Economic*, Vol.47, No.3, pp. 67-72, July 15.

Kumar, Manish, Kalpana Kumari, A.L. Ramanathan and Rajinder Saxena (2007), "A comparative evaluation of ground water suitability for irrigation and drinking purposes in two intensively cultivated districts of Punjab", *Environmental Geology*, Vol. 53, No. 5, pp. 89-93, Nov. 3.

Kumar, Surender and Nandini Chandra (2010), "Sustainability concern in Indian Agriculture: An Analysis of Subsidies and Investment", *Economic and Political Weekly*, Vol. 3, No. 2, pp. 123-125, Nov. 16.

Mahasevan, Renuka (1993), "Productivity growth in Indian agriculture: The sole of globalization and economic reform", *Asia-Pacific Development Journal*, Vol. 10, No. 2, pp. 67-80, Feb 13.

Majumdar, H.H. (1993), "Fertilizer subsidy in India - who benefits?", *Economic and Political Weekly*, Vol. 67, No. 11, pp. 86-93, April 18.

Malik, Anjali (2004), "Post Subsidy removal Fertilizer Consumption-A study of farmer perspective", *Fertilizers News*, Vol. 49, No.9, pp. 59-64, April 20.

Malik, R.P.S. (2005), "Indian Agriculture: Recent Performance and Prospects in the Wake of Globalization", *Economic and Political Weekly*, Vol. 4, No. 3, pp. 170-174, July 13.

Mitra, Ajit Kumar (1982), "Subsidy as an instrument for improving the economic conditions of the marginal and small size category farmers: A case study of the district of Ganjam Odisha, India", *Journal of Agricultural Economic*, Vol. 37, No. 3, pp. 278, July-Sept.

Saibal Kar and Beladi Hamid (2007), "Protectionary bias in agriculture : A pure economic argument", *Economic and Political Weekly*, Vol. 63, No. 1, pp. 160-164, June 15.

Maurice, Landes and Mary Burfisher (2008), "Investing in India's Agriculture markets: A source of growth and equity?", Conference paper presented at the 11th annual conference on global economic analysis, Helsinki, Finland.

Mukherji, Aditi, Tushaar Shah and Shilp Verma (1990)," Electricity reforms and their impact on ground water use in states of Gujarat, West Bengal and Uttrakhand, India", *Economic and Political Weekly*, Vol. 35, No. 2, pp. 145-150, Feb.15.

Namboodiri, N V (1982), "The Rational of Input Subsidies: The case of Fertilizer", *Indian Journal of Agricultural Economic*, Vol. 37, No. 3, P 280, Jan. 6.

Narayanamoorthy, A. (1997), "Impact of Electricity Tariff Policies on the Use of Electricity and Groundwater: Arguments and Facts", *Artha Vijnana*, Vol. XXXIX No.3. pp. 323-340, September 4.

National Bank for Agriculture and Rural Development (NABARD) 1991, Report of the working group on lift irrigation schemes, NABARD, Mumbai.

Orden David (2003), "Agricultural Policies in India : Producer Support estimates 1985-2002", *Economic and Political Weekly*, Vol. 53, No. II, pp. 73-81, Feb. 18.

Pachauri, R.K. (2006), "Power Politics : no light at the end of the H. John (2006), "Should agricultural subsidies be banned?", *Economic and Political Weekly*", Vol. 27, No. 5, pp. 71-78, April 23.

Pal, Parthaprathim (2001), "Current WTO negotiations on domestic subsidies in agriculture : Implications for India", *Economic and Political Weekly*, Vol. 61, No. 9, pp. 81-92, August 17.

Pal Parthapratim (2005), "Current WTO negotiations on domestic subsidies in Agriculture: Implications for India", Working Paper No. 177, Dec 3.

Pradhan, P.N., S. Jena, A.K. Mitra (2000), "Growth of fertilizer consumption in Odisha-a district-wise analysis", *Economic and Political Weekly*, Vol. 7, No. 3, pp. 23-30, March 25.

Raj. K.N. and Shriniwas (1984), "Cost benefit analysis of price support and input subsidy to achieve wheat production target in India", *Economic and Political Weekly*, Vol.38, No. 12, pp. 809-812, March 19.

Reddy, K S. (2005), "Who Benefits from Agricultural Subsidises? The case of Andhra Pradesh", *Economic and Political Weekly*, Vol. 25, No. 1, pp. 64-72, June 12.

Reema (2010), "India changes Agricultural subsidies to promote soil health", *Wall Street Journal*, Vol. 4, No. 3, pp. 56-60, Feb. 24.

Regina Birner, Nethra Palaniswamy and Neeru Sharma (1990), "Understanding the Political Economy of Agricultural Subsidy Reform -The Case of Electricity

Subsidies for Groundwater Irrigation in India", *Economic and Political Weekly*Vol. 25, No. 41, PP1727-1732, Oct. 13.

Sant, Girish (1996), "Beneficiaries of the IPS subsidy and the impact of tariff-hike", *Economic and Political Weekly*, Vol. 12, No. 1, pp. 494-496, December 21.

Schwartz, Gerd and Clements Bonedict (1997), "Government subsidies," *Journal of Economic Surveys*, Vol. 13, No. 2, pp. 119-148, Dec. 16.

Sengupta, T.K. (2004), "Global trade of fertilizers – an Indian perspective", *Economic and Political Weekly*, Vol. 49, No. 9, pp. 55-57, Feb. 11.

Sharma, V.K. (1982), "Impact of agricultural subsidies on Nation Income and agricultural production", *Economic and Political Weekly*, Vol.47, No. 7, pp. 66-71, August 18.

Sharma, Vijay Paul, Hrima Thaker (2010), "Fertiliser Subsidy in India: Who Are the Beneficiaries?", *Economic and Political Weekly*, Vol. 26. No. 12, pp. 45-52, March 20.

Sharmrao, Tanpure Sambhaji (2011), "A study of fertilizer policy in India", *International Journal of Agriculture Sciences*", Vol. 3, No. 3, pp. 145-149, Dec. 9.

Sidhu, Jagrup Singh and D.S. Sidhu (1985), "Price-Support versus fertilizer subsidy: An evaluation", *Economic and Political Weekly*, Vol.20, No. No.13, pp. A-17 to A-22, March 30, 1985.

Singh, Amarjit (1998), "Economic and Policy Dimensions of Irrational use of Electricity in Agriculture in India", *Indian Journal of Economic*, Vol. 79, No. 313, pp.187-193, Oct. 1.

Singh, Gajendra and Mani, Indra (2008), "Influence of legislation/subsidies to help agriculture and agricultural mechanization on the market of agricultural machinery in India", *Economic and Political Weekly*", Vol.38, No. 3, pp. 256-160, Aug. 23.

Singh, Richa (2004), "Equity in fertilizer subsidy distribution", *Economic and Political Weekly* Vol. 39, No. 3, pp. 295-300, Nov. 15.

Singh, Shivcharan (2008), "Subsidies never reach farmers", *Indian Express* July 28.

Sirohi, A.S. (1984), "Impact of Agricultral subsidies and Procurement Prices on Production and income distribution in India", *Indian Journal of Agricultural Economic*, Vol. 39, No. 4, pp. 563-585, October 18.

Srivastova, D.K., C. Bhujango Rao, Pinaki Chakraborty and T.S. Rangamannar (2003), "Budgetary Subsidies in India: Subsidising social and economic services", National Institute of Public Finance and Policy, New Delhi, March 12.

Sud, Surinder (2009), "Direct transfer of fertiliser subsidy to farmers misconceived: Study", *Economic and Political Weekly*, Vol. 14, No. 6, pp. 56-61, Nov. 23.

Virk, Bikram Singh (2008), "Farm prices and subsidies need a relook", *The Tribune,* June 23.

Yogesh, Bandhu (2009), "Trade Competitiveness, Subsidies and Barriers to Trade: Implication for Indian agriculture", *Economic and Political Weekly*, Vol. 4, No. 3, pp. 145-150, Nov. 4.

www.ingramcontent.com/pod-product-compliance
Ingram Content Group UK Ltd.
Pitfield, Milton Keynes, MK11 3LW, UK
UKHW021533300726
14060UKWH00011B/478

9 789390 384044